**A gift note from
Daniel J Perkins:**

Happy Father's Day Dad! Hope you
enjoy this book! Love, Dan

Included with Classic John Deere Two-Cylinder Tractors: History,
Models, Variations & Specifications 1918-1960 (Tractor Legacy
Series)

CLASSIC
JOHN DEERE
TWO-CYLINDER
TRACTORS

John Dietz
and
Jeff Hackett

Voyageur Press

First published in 2008 by Voyageur Press, an imprint of MBI Publishing Company, 400 First Avenue North, Suite 300, Minneapolis, MN 55401 USA

Voyageur Press titles are also available at discounts in bulk quantity for industrial or sales-promotional use. For details write to Special Sales Manager at MBI Publishing Company, 400 First Avenue North, Suite 300, Minneapolis, MN 55401 USA.

To find out more about our books, join us online at www.voyageurpress.com.

Library of Congress Cataloging-in-Publication Data

Dietz, John, 1946-
  Classic John Deere two-cylinder tractors : history, models, variations & specifications 1918-1960 / by John Dietz ; photographs by Jeff Hackett. -- 1. ed.
      p. cm.
  Includes index.
  ISBN 978-0-7603-3197-2 (hb plc)
  1. John Deere tractors--United States--History. 2. Farm tractors--United States--History. I. Hackett, Jeff. II. Title.
  TL233.6.J64D543 2008
  629.225'2--dc22
                              2008004097

ISBN-13: 978-0-7603-3197-2

Editor: Amy Glaser
Designer: Elly Gosso
Cover designed by Koechel-Peterson and Associates, Inc., Minneapolis, MN

Printed in China

CONTENTS

ACKNOWLEDGMENTS

Most of the material in this history has been drawn from other published material, either in print or online. Several individuals have provided personal recollections. Others have read portions of the manuscript. Collectors should rely on authoritative sources, such as Deere & Company Archives, the John Deere Collectors Center, Two-Cylinder Club, and *Green Magazine*. On a personal basis, I want to offer special thanks to: Larry Baker, Ken Burns, Neil Dahlstrom, Peter Easterlund, Kenny Earman, Elmer Friesen, Arvin Hagen, Travis Jorde, Kenny Kass, Jack Kreeger, Harlan Kruger, Charles Lindstrom, Mike Mack, Malcolm McIntyre, Gerald Mortensen, Mike Ostrander, Steve Ridenour, Robert Serlet, Greg Stephen, Gene Tencza, Gary Uken, and Albert Ulrich.

INTRODUCTION

John Deere two-cylinder tractors were manufactured in three locations between 1924 and 1960. The history can be approached as a tale of three factories in which one of them was clearly dominant. Go back in time to the year 1850. The pristine landscapes of North America are becoming tracked with steel. Fences are being built. Factories are belching black smoke. A teeming nation of men and boys tread daily from home to work and work to home. Boundless land has been surveyed and staked into millions of parcels. Oxen break sod and many hands are needed. Land supplies the city; city supplies the land. There is a new thing called the telegraph and an even newer thing called the telephone. There is steam, which produces great power, drives enormous machines in factories, and drives locomotives on steel. Men are dreaming of new things, of easier living, and of a day when a small farm could feed more than two families and a team of horses. "If only I had the power of steam on my farm, what a wonder it would be." It took a lifetime to achieve, about 70 years, but the new day dawned.

The first steam-powered traction engine was built in 1841 by Robert Ransomes of Ipswich, England. In 1876, the Otto internal combustion engine was patented. In the late 1870s, Jerome Increase Case and partners began building steam traction engines in Racine, Wisconsin. In 1887, John Charter invented the first liquid fuel traction engine in Sterling, Illinois. In 1892, the first gasoline-powered traction engine sparked new hope.

At the University of Wisconsin in Madison, Charles W. Hart and Charles H. Parr began developing gas engines while studying mechanical engineering. They formed the Hart-Parr Gasoline Engine Company in 1897 and moved it to Hart's hometown of Charles City, Iowa, in 1900. They erected the first factory in the United States that was dedicated to producing gas traction engines. They are also credited with coining the word "tractor" for machines that had previously been called gas traction engines. Their first tractor, Hart-Parr No. 1, was built in 1901. However, many factories were on the same trail. By 1908, the first North American tractor trials were organized in Winnipeg, Canada. Three years later, the first United States tractor demonstration was held in Omaha, Nebraska.

As of 1910, a thousand tractors had been built. America had more than 24 million horses and mules, which was about three to four animals for each farm. Farms were growing wheat, corn, and cotton; raising livestock; producing dairy and other products; and relying on steam-powered engines, draft animals, or their own strength for power. Implements included plows, disks, harrows, planters, cultivators, mowers, and reapers. Plowing required the most power. As of 1910, companies were building around 4,000 steam tractors a year, but growth in animal horsepower was faster. The typical steam tractor or traction engine weighed 20,000 pounds, it was expensive, and it was practical only for a very large farm. Millions of regular and small farms needed traction power from a machine to replace the horses and mules, but it had to be affordable and reliable.

As of 1915, 14 companies in Illinois were building traction machines for power to plow a field in place of horses. Deere & Company, an established company in Moline, was developing its own tractor but it certainly wasn't in the lead.

Continued on page 11

Deere & Company, the world's largest manufacturer of agricultural equipment, traces its tractor lineage back to an Iowa inventor, John Froehlich. He developed the first successful gasoline-powered farm tractor.

In 1841, the first steam-powered traction engine was introduced by Robert Ransomes of Ipswich, England. By the late 1870s, Jerome Increase Case and partners were building steam traction engines in Racine, Wisconsin.

The early traction engines were expensive, they sometimes needed an hour to build up steam, and they were mostly useless for farm work. Smaller and lighter steam engines were developed, but they were still costly beasts with large appetites for clean water and coal. Farmers continued to rely on the horse, mule, and ox for power to pull the moldboard plow.

In 1845, Johannes Heinrich Froehlich emigrated from Germany and arrived in America. He married, bought an Iowa farmstead about 100 miles north-northwest of Moline, and started a family. John Froehlich was the first of nine children. Born on November 24, 1849, he grew up to be a farmer but his interests went beyond farming. Soon, he was operating a grain elevator at the village of Froehlich and each fall he ran a threshing operation about 500 miles northwest, near Langford, South Dakota. Fascinated by machinery, he became very familiar with the steam-powered traction machines and their deficiencies.

In 1890, Froehlich bought a new kind of internal combustion engine from the Van Duzen Gas and Gasoline Engine Works of Cincinnati, Ohio, to power his grain elevator. He also began experimenting with making a gasoline-powered traction engine. In 1892, he mounted the single-cylinder Van Duzen on a Robinson chassis. Using a traction mechanism of his own design, Froehlich's machine could move forward and backward, as well as operate a threshing machine with a long belt. For the 1892 harvest, Froehlich and assistant William Mann hooked their new traction engine to a J. I. Case thresher and harvested 72,000 bushels of grain in 52 days and worked in temperatures ranging from 100 degrees Fahrenheit to below zero.

This 1915 John Deere penny post card encouraged readers to accept a free copy of a farm paper, *The Furrow*.

Inspired by success, Froehlich and investors from Waterloo formed the Waterloo Gasoline Traction Engine Company in 1893. The company built four tractors and two were returned by unhappy customers. In 1895, the investors reorganized as the Waterloo Gasoline Engine Company and focused on building gasoline engines for stationary applications, such as powering grain elevators, and Froehlich decided to leave the firm. Without Froehlich, the Waterloo Gasoline Engine Company continued trying to develop a gasoline engine for transportation uses. In 1902, the company reorganized and merged with the Davis

The John Deere Waterloo Works had been expanded and modernized by the time of this 1924 photo.

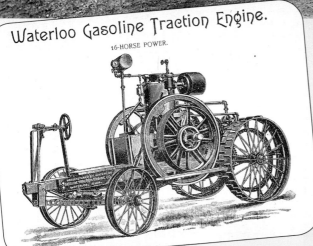

Waterloo Gasoline Traction Engine.

16-HORSE POWER.

John Froehlich's traction machine was designed around 1894 for the Waterloo Gasoline Traction Engine Company.

Engine Company as the Waterloo Tractor Works to manufacture stationary engines and automobiles. Two years later, the merger was over and the Waterloo Gasoline Engine Company was on its own again. However, it had a new product: the 1904 Waterloo

Boy engine. It was reliable, cheaper to operate, relatively lightweight, and stationary. Seven years later, it had an improved engine and traction mechanism.

It was called the Waterloo Boy engine and it burned kerosene. The little factory worked with various designs to transfer power to a steerable set of wheels. It would have traction for pulling and a belt for powering separate equipment. One design followed another in 1912 and 1913. The first Waterloo Boy Model L, a three-wheel tractor, was shipped to California on January 26, 1914.

According to historian J. R. Hobbs, "Serial production of the model L Waterloo Boy tractor began with serial number 1000 (in 1914), though it is known that L tractors, along with a larger model, the H, were produced in limited quantities before this time. Serial production of the L was confined to two tractors, number 1000 and number 1001, superseded by the revised model LA.[1]

[1] The John Deere Unstyled Letter Series, pub. 2000, by JR Hobbs, @Green Magazine, page 22

This is an artist's rendition of a Waterloo Boy, as designed in November 1914 for the Waterloo Gasoline Engine Company. Notice the chain steering.

The Model LA, a four-wheel version, was shipped the same day to California and 27 more were shipped by April 21, 1915, in configurations with three and four wheels. None of these early models is known to exist today. For the following year, Waterloo offered the Model R. When Model R was discontinued in 1918, the Waterloo manufacturer was already well established with a different, popular tractor model. A success story had been written: The larger Waterloo Boy Model N tractor was being built and the company had decided to abandon stationary engines.

Meanwhile, over in Moline, Deere & Company had factories producing very successful lines of farm implements. Instead of tooling up a factory and introducing another tractor, Deere & Company made a buyout offer that was too good to refuse. In 1918, Deere & Company bought out the Waterloo manufacturer for $2.2 million and went into tractor production with a fully equipped factory that could build a popular tractor.

In 1924, workers began building a new tractor in the factory that Froehlich had started. This was called the John Deere

Model D. It was the first time the words "John Deere" appeared in tractor paint, and it was designed from the ground up by Deere & Company engineers. It wasn't a bad way to start. The Model D stayed in production for 30 years. New buildings and tractor assembly lines were added at Waterloo in the 1930s and later. By the end of the two-cylinder tractor era in 1960, the Waterloo factory complex had built approximately 1,298,000 tractors and had gone through several expansions. It continues to serve Deere & Company today.

What about John Froehlich? He married Kathryn Bickel and had two sons and two daughters. For awhile, Froehlich built engines for Novelty Iron Works in Dubuque. In 1910, he was working for his brother Gottlieb Froehlich when he became vice president of Henderson-Froehlich Manufacturing. He moved to St. Paul, Minnesota, to work as an investment counselor until the stock market crash of 1929. He spent his last years with little money and lived with his daughter Jenetie in St. Paul. He died in 1933. In 1991, John Froehlich was inducted into the Iowa Inventors Hall of Fame.

A 1935 Model D is shown at work threshing in a field.

Continued from page 7

The big breakthrough came in 1917, over in Dearborn, Michigan. Henry Ford had begun experimental work on a small, gasoline-powered tractor a decade earlier and he referred to it as an "automobile plow." His efforts were delayed, but eventually the Fordson Model F four-cylinder, three-speed tractor was ready for sale in 1917. It was revolutionary in its scaled-down design and was a great success. Ford did not invent the tractor or the automobile; however, he did make both affordable and reliable for American farmers.

According to the U.S. Department of Agriculture reports, it took 40 to 50 labor hours to produce 100 bushels of wheat on five acres with a gang plow, seeder, harrow, binder, thresher, wagons, and horses in the 1890s. By 1930, a three-bottom gang plow, tractor, 10-foot tandem disk, harrow, 12-foot combine, and truck could produce the same crop in 15 to 20 hours.

When the barriers were broken and mechanical power was tamed for ordinary farms, tractor sales began to soar. Approximately 250,000 tractors were built in

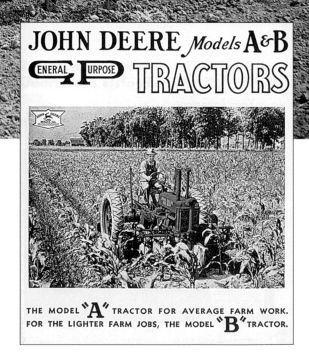

ABOVE: This farmer is cultivating a corn field on his Model A.

RIGHT: This booklet provided a farmer with all the available options for the A and B tractors, as well as endorsements from fellow farmers who owned one of these machines.

the United States in 1920. The Great Depression and World War II slashed tractor sales, but they rose to 550,000 in 1949 and peaked at 564,000 in 1951. By 1960, an estimated five million tractors had been built for American farms and only three million horses and mules remained as power sources.

The availability of millions of tractors by many companies and from many factories fundamentally changed the nature of farm work. It expressed the dream of a developing country and altered the structure of rural America. It freed millions to leave the farm to work in manufacturing and service industries. It had an economic

SPECIFICATIONS
JOHN DEERE
MODEL "L" TRACTOR

JOHN DEERE
Model "L"
TRACTOR

CAPACITY—Handles load ordinarily pulled by 2 horses. (Daily work capacity is much greater than that of 2 horses because of constant speed and faster working speeds of tractor.)

SPEEDS—First, 2-1/2 miles per hour; Second, 3-3/4 miles per hour; Third, 6 miles per hour; Reverse, 3-3/4 miles per hour.

BELT PULLEY—Extra equipment. 6-3/4" diameter with 4-1/2" face (6-5/8" face optional.) Speed, 1,480 revolutions per minute.

BELT SPEED—2,615 feet per minute.

ENGINE—2-cylinder, vertical. Rated speed, 1,550 revolutions per minute controlled by variable speed governor. Bore, 3-1/4 inches. Stroke, 4 inches. Ignition, high-tension magneto with enclosed automatic impulse starter. Carburetor, feed load jet with adjustable idle jet. Air cleaner, oil-wash type. Lubrication, force-feed pressure system. Oil capacity, 2 quarts. Cooling, thermo-siphon. Water capacity—2-1/2 gallons. Fuel Tank Capacity—6 gallons. (Fuel, regular gasoline.)

CLUTCH—Single dry plate, automotive-type, foot operated.

TRANSMISSION—Selective-type spur gears, alloy steel, forged, cut, and heat-treated.

FINAL DRIVE—Enclosed gear-driven, running in oil.

REAR WHEEL TREAD—Adjustable, 36 inches and 42 inches, center to center of wheels. Front wheel tread is fixed—rear wheel tracks cover front wheel tracks in both tread positions. An extra-wide tread of 54 inches can be secured by reversing the rear wheels, but the 36- and 42-inch treads meet practically all farming needs.

TURNING RADIUS—7 feet.

DRAWBAR—Swing type. Vertical adjustments, 12, 14, and 16 inches above ground. Horizontal adjustment, 18-1/2 inches.

DIMENSIONS—Wheelbase, 61 inches. Over-all width, 49 inches. Over-all length, 91 inches. Height, top of radiator, 50 inches.

BEARINGS—22 Timken tapered roller bearings. 4 precision ball bearings. 1 precision straight roller bearing. Alemite grease-gun lubrication.

WHEELS—Disk-type with low-pressure rubber tires all around. Front tires, 4.00 x 15. Rear tires, 6.00 x 22.

SHIPPING WEIGHT—1,570 pounds.

Note—Specifications are subject to change without notice.

The Lightweight, Economical Tractor That Handles All Work Ordinarily Done with a Team of Horses.... with Complete Integral Equipment for Plowing, Planting, Cultivating, and Mowing.

ABOVE: A Model L, with a long day ahead of it, pulls a one-row potato planter around 1939. It was the first John Deere tractor with a vertical two-cylinder engine and a foot clutch. This one also had the Dreyfuss styling treatment.

LEFT: This booklet spoke the praises of the Model L and provided information for farmers who were looking for a smaller tractor for the farm.

and social impact that rivaled the automobile, truck, train, and the airplane. The impact was worldwide. It became a defining symbol of the developed world in the twenty-first century.

One factory in the heartland of that modern development was located in a small, northeastern Iowa city called Waterloo. Another factory was in a town in northwest Illinois named Moline. Dubuque was located in between the two cites and people in Dubuque were thinking about other things. This is the tale of one company and three factories.

CHAPTER 1

IN THE BEGINNING
WATERLOO BOY MODEL R

The first Waterloo Boy Model R was shipped on August 8, 1914. According to archives, it appears that only 46 tractors were built by the end of 1915. The last recorded shipping date was December 18, 1917. The production run had many minor variations and came to approximately 9,300 Model R Waterloo Boy tractors. It is estimated that less than 100 still exist, or less than 1 percent of production. More than any other single tractor model, the Waterloo Boy Model R is the bedrock of modern tractor production. There's more to the story, of course, but fundamentally it was the first tractor anywhere that went into full scale commercial production.[1]

The Waterloo Boy R represented a breakthrough in technology. It was very reliable compared to other early tractors. It was easy to operate, rugged, priced right, and was powered by kerosene. The two cylinders fired at a 180-degree firing order and produced an engine that was powerful and compact. The tractor was lighter than most alternatives in 1914 to 1916, and at $750, it was less expensive. It also burned kerosene, which was abundant and inexpensive. The tractor was rated for operating two 14-inch plows, but it could pull a third plow in most field conditions. It had a gearbox with two gears: forward or reverse. In forward gear, it chugged along at 2.5 miles per hour, which is slow by later standards. Yet, it was faster than any team of horses doing the same job.

The original Model R engine block was bored to 5.5 inches for each cylinder. The pistons had a 6-inch stroke. The bore was increased twice during the production run to give the engine more power. The bore was 6 inches for R engines in 1915 and 1916. After 1917, the bore was widened to 6.5 inches. The piston stroke stayed a constant 7 inches. The rated rpm also increased from 700 in the first models to 750 rpm in the later engines. The third generation Waterloo Boy R engine had a 465 cubic inch capacity. It wasn't fast, but it had power. The entire Waterloo Boy R tipped the scale at more than 5,300 pounds of steel, iron, brass, aluminum, copper, tin, and glass. With steel wheels on loose soil, it could tirelessly and faithfully pull two or three 14-inch plows.

From the outset, the Deere engineers were always improving the product. Components in the production system were almost constantly altered, tweaked, or changed. It would not be unique to Waterloo Boy.

---

[1] Official records for Waterloo Boy and later two-cylinder John Deere tractors are available through Two-Cylinder® Club, at Grundy Center, Iowa. Two-Cylinder® Club is a nonprofit service, educational, and recreational organization. It extends membership to anyone interested in preservation of John Deere tractors and implements.

OPPOSITE: A fully restored, picture-perfect 1916 Waterloo Boy Model R owned by Robert Warriner of Montrose, Pennsylvania. It was restored by Kenny Kass of Dunkerton, Iowa.

Here is a side view of the Model R, which is ready for the next parade. The steel wheels are covered with rubber belting to protect the wheels and the operator.

Feedback was received and changes were made to improve the already popular product. Virtually every Deere two-cylinder tractor model was modified many times during the production cycle.

The Waterloo Boy R, however, may have been the most modified or revised tractor in history. Twelve major changes are recorded in 40 months of shipping. Waterloo designated each change with a second letter after the R, thus proceeding alphabetically from the RA to the RM. Most of the changes were associated with valves and magnetos.

For instance, before Deere & Company took ownership, the Waterloo company built 39 three-wheel tricycle versions of the Model R tractor between December 15, 1914, and August 31, 1917. They were destined to work in orchards and vegetable crops along the West Coast and became known as the California Special[2]. Not one California Special is known to exist

2 Ibid., page 23

A 1919 Waterloo Boy Model N tractor built in late 1918 is shown at the Homer Kolb farm in Phoenixville, Pennsylvania. It was fully restored to this pristine condition by Kenny Kass around 1995.

today. Within the California Specials, according to Two-Cylinder Club historian J. R. Hobbs, there were variations. Differences included one or two front wheels, an optional integral two-bottom plow, and a more maneuverable short wheelbase version.

During 1914 to 1918, while World War I was raging in Europe, the American farm economy was relatively prosperous. Demand increased for exported farm products, including the new tractors. Farmers in Britain were also looking for a traction machine that could replace teams of horses. Waterloo Boy R fit the niche. During the war years, Deere & Company exported about 4,000 Waterloo Boy tractors to Britain. They were repainted, given a new name of Overtime, and were sold primarily to farms in England and Ireland. Other export destinations included France, Greece, Denmark, and South Africa.

In just a few years, the Waterloo Boy Model R could be found at work in most farming areas of North America. Many were still being used in the 1930s. However, by the time the last Waterloo Boy Model R

rolled out of the factory, a new, sleek, and smaller tractor already was coming off a new high-speed assembly line in Detroit. Within a short amount of time, there were better tractors from several companies. By 1930, the Model R must have seemed quaint. Gradually, the machines that had been the foundation stones of modern farming were parked somewhere and mostly forgotten. World War II brought a huge demand for iron, copper, and other metals to support the war effort. By 1945, nearly every Waterloo Boy Model R had become scrap metal. Most were recycled into material for war supplies. Only a few were still tucked away to be discovered again and regarded as new-found treasure decades later.

A Word to Collectors

The Waterloo Gasoline Engine Company manufactured stationary two-cylinder Waterloo Boy engines for several years before it introduced the

A Waterloo Boy Model R, style H, is hard at work with a three-bottom plow.

Waterloo Boy Model R tractor. The tractor had one flywheel and most of the stationary engines had two flywheels. Stationary engines with the single flywheel were very similar to the tractor engine with a single flywheel. It also appears that the company intermingled serial numbers between the two types of engines, which has led to some confusion for modern collectors.

Model N

The Waterloo Gasoline Engine Company shipped its first Model N tractor on January 3, 1918; just two weeks after the last Model R had shipped out. The initial price was $1,050. The last Model N was shipped on October 15, 1924. Approximately 20,500 tractors in this series were built during that time frame, which equals about 100 tractors for each working day at Waterloo. No one knows how many survive today, but a fair estimate seems to be about 1 percent.

The Model N addressed the biggest complaint about the Model R. This tractor had an enclosed two-speed transmission. It could move in high gear at about 3 miles per hour, which gave the operator more flexibility in tackling field chores. It retained the 365-cubic-inch engine that was first used on the R to the end of the N's production cycle in 1924.

On March 14, 1918, Deere & Company purchased the Waterloo Gasoline Engine Company. Deere took over the management and kept Waterloo Boy Model N in production.

Big and small refinements continued in the Model N series. Engineers soon developed a larger radiator. In March 1919, the fuel tank was raised 3 inches. In June, it was raised another 3 inches. The serial number plate and the crankshaft counterweight were both moved. The fan and water pump were combined into a single unit. New sets of decals were created for the Model N, and the painting scheme was revised in January 1920. The Waterloo Boy Model N became "John Deere green," except for the red hubcaps and red and yellow accents at a few strategic points.

The front view of the Model N clearly shows the chain steering system. The rear wheel lugs were made of angle iron and bolted at an angle across the wheel rim.

A major change was introduced in February 1920 as engineers brought in a better steering system. Until now, operators steering the Waterloo Boy were actually pivoting the front axle using heavy chains and a roller mechanism. Eventually, the chains and rollers wore down. To turn the tractor, it would gradually require more and more turns of the steering wheel. Deere & Company replaced the chains-and-roller system with a steering system that is similar to one used on automobiles, and it was called "auto steer."

A Waterloo Boy N with auto steer made history for another reason in the spring of 1920. It was the first tractor tested under a new tractor testing law in Nebraska. The Nebraska Tractor Tests became the

This is a full-page ad for the Waterloo Boy in *Farm Implements* magazine.

This Waterloo Boy Model N two-speed pulls a big disker and is operated by one of the John Deere engineers. The Model N series went into production in January 1918.

industry standard for rating and comparing tractors. Nebraska Test No. 1 was performed on the Waterloo Boy N on March 31, 1920.

In August and September 1920, Waterloo equipped 24 N tractors with a special air filtering system (the clarifier) for dusty field conditions. The 24 California Clarifier[3] tractors served the rapidly growing, commercial vegetable industry in California and Oregon, as had the previous California Special. Only one California Special is known to exist today.

A change to the Waterloo Boy N was implemented on September 30, 1920. Beginning with serial number 28094, the Waterloo Boy was given a riveted frame rather than a bolted frame.

By the end of World War I on November 11, 1918, the new age of cars, trucks, and tractors was well

underway. Farmers were convinced that tractors should replace horses, even if they couldn't afford a tractor yet. Waterloo Boy was only one of several tractors offered and it was well behind the best-selling tractors. Market leaders for tractors in 1924 were International Harvester (IH) and Ford. Henry Ford, who had tinkered with steam and gasoline tractors prior to achieving his success with automobile production, introduced a small, inexpensive model that he called the Fordson during World War I. At $625, it was immediately popular. After the war, farm prices plunged and tractor sales fell sharply around 1920 and 1921. In 1922, Ford slashed the Fordson price to $395 and latched onto more than 40 percent of market share.

Only International Harvester kept pace with Ford. IH matched the price cut and introduced a significant improvement—the first power take-off—in 1922. This

[3] Ibid., page 31

The Waterloo Boy Model R, shown here, was in production at the Deere & Company bought the Waterloo Boy Company. Production of the R discontinued after 1918 when it was replaced by the Model N.

device allowed a direct drive to implements from the tractor engine. The power take-off (PTO) quickly became a standard feature on all tractors and implement makers began re-designing equipment to take advantage of this innovation. Generally International Harvester tractors captured around 25 percent of tractor sales in the 1920s. Other early 1920s tractor makers included Case, Allis Chalmers, and Minneapolis-Moline.[4]

Waterloo Boy wasn't really competitive by 1923. Its sales were about 4 to 6 percent of the market. If Deere & Company was going to remain in the tractor business, it needed to build a lower cost tractor that matched or exceeded the performance of the competitors. It did.

Engineers are hard at work and hunched over their drafting boards at the Waterloo factory in 1920.

[4] Economic History of Tractors in the United States, William J. White, eh.net/encyclopedia/article/white.tractors.history.us

WATERLOO FACTORY BIG DADDY TRACTORS
MODELS D, R, 80, 820, 830

Model D

The John Deere Waterloo tractor factory was at the threshold of a new era in farming during the early 1920s. It underwent several expansions through 1960 and it introduced many lines of tractors tailored to the needs of different farms. Within this text, the production is divided into four chapters that are roughly arranged by family line and sequence. This chapter covers the largest tractors.

In the early 1920s, tractor engineers at several companies were taking big steps in technology. Ford slashed its price for the Fordson from $625 to $395 in 1922. International Harvester matched Ford; in 1922, it put the first major tractor innovation on the market: the power take-off. A shaft turned by the rotating crankshaft that allowed implements to be driven directly by the tractor engine. International Harvester Company (IHC) went a step further in 1925 and introduced the Farmall, the first general-purpose tractor. It had small front wheels, high clearance, and minimal weight. In addition to pulling plows, it could cultivate rows.

The 1923 John Deere Model D was a big step forward in technology compared to the Waterloo Boy. The Model D became a huge success and stayed in production until 1953. The heart of the D was Waterloo's sturdy, gasoline-fired, two-cylinder engine. During the 1940s, engineers developed a diesel-fired, two-cylinder engine that could outperform the Model D gasoline engine. The company introduced its new Model R diesel tractor in 1949 while the Model D was still in production. In 1953, it was time to change the nomenclature from alphabetical to numerical names.

The tractor families featured in this chapter were large and well suited to mid-size and larger Midwestern farms. They relied on a low center of gravity, serious weight from lots of iron, and a sturdy two-cylinder engine for tremendous pulling power. Several wheel types were offered, but the basic tractor had only one configuration.

Model D and Model R families were built with a simple, straight, beam-axle design that became known as standard tread. The front wheel tread was as wide as the rear tread. The front axle pivoted to steer the Waterloo Boy. For Model D and Model R tractors, the front axle was fixed in place. Engineers equipped the tractor with running gears, which were used on automobiles and steam engines, to turn the front wheels rather than pivot the axle. Later, row crop tractors had new configurations.

Waterloo built approximately 160,000 Model D tractors. The tractor was so successful and basic that

OPPOSITE: This classic 1925 John Deere Spoker D tractor is owned and was restored by Bruce Sniffin of Vernon, Connecticut. It was used on a Connecticut farm for a few decades before its restoration. Sniffin won a local award in 1999 for his reconstruction project. The original fenders and wheels were replaced, and the engine was rebuilt.

Deere & Company kept building it longer than any other tractor it has ever produced. The production life of many models was less than five years. However, in all fairness it must be noted that components of the Model D were often improved and the upgrades give it great diversity. A specialized Model D collection can be quite large.

Approximately 39,000 powerful, standard-tread, two-cylinder diesel tractors were built at Waterloo in 12 years. By 1960, the diesel-fueled tractor family had four name changes. A few of the old standard-tread kerosene or diesel tractors still are at work today. The Model R also is a popular big-power tractor at tractor pulling competitions.

A word must be said about assigning nomenclature. In the days of traction engines, engineers designated new designs alphabetically as Model A, B, C, and so on. When a model's component was changed, it was appropriate to add a second letter. Hence, the Waterloo Boy Model L and Model LA. After World War II, a numerical sequence of several numbers became favored. In 1955, the Model

R morphed into the Model 80, and other tractor families followed the pattern. Two years later, the big diesel got a third digit in the name that signaled significant improvements to the basic family. The Model 80 became the 820 and it was followed by Model 830. Deere & Company was gearing up for a New Generation of tractors for 1960. These models had four cylinders and four digits in the name, but that's another story.

The Model D was an instant success in its early years. It sold itself on the basis of performance in the field, economy, simplicity, and accessibility. If you farmed with one or two big horse teams in the Midwest, the Model D was a very good option. It was a heavy duty tractor that was built to pull a three-bottom plow or other implements on big, wide open fields. It also could operate a wide belt to provide tractor power for other farm work. Sales were strong for most of its production life except for the middle of the Great Depression, World War II, and the early 1950s.

The spoked rear wheel on this 1925 Spoker D has the classic cast-iron lugs bolted onto the steel rims. They had great grip in the fields, but were hard on operators if they couldn't dig into the dirt. This tractor is owned by Grant Zook and Harold Gahman of Souderton, Pennsylvania.

Here is a close-up of a spoked flywheel on the 1925 Model D. This spoked flywheel was made for only two years.

This 1925 Spoker D parked in a field of daisies and dandelions is owned by Grant Zook and Harold Gahman of Souderton, Pennsylvania. Almost everything on the tractor is original except the hood. It came from Iowa and was completely restored in 2004 by Zook and Gahman.

Within the Model D family, there are early, mid-, and late unstyled D tractors from 1923 through 1939. Waterloo manufactured 23,000 units in the early series, which spanned from 1923 through 1927. In the next three model years, 1928 through 1930, the Waterloo factory built 56,000 tractors. Thanks to a half-inch-wider cylinder bore, the mid-D had slightly more than 28 horsepower on the drawbar and 36 horsepower on the belt when it was tested in Nebraska. The late, unstyled D was introduced for 1931. The engine bore and stroke stayed the same, but the engine was all new from the crankshaft forward. The late D series ran for nine model years through 1939 with a total production of 33,000 units.

The early D came standard for $1,112 with spade-like, four-inch lugs on the four steel wheels, a swing

A Model D in snow is ready for its restoration next spring. This photograph was taken near New Milford, Connecticut.

drawbar, operator's platform, and fenders. By 1926, popular options included a $3 muffler, $1 steering wheel handle, $2 radiator curtain, six-inch rear wheel extension rims for $31, and a power shaft assembly for $38. The mid-D basic price with a power shaft assembly increased to $1,119. A late D price list offers the basic tractor with huge five-inch spade lugs on the steel wheels for $1,125. The spoke wheel option for new pneumatic tires increased the base price to $1,400.

Mid- and late Model D tractors had many options. Trying to find them all today is one of the joys or challenges for modern collectors. Wheel options are one example. It appears that the first Deere tractor equipped with rubber tires for agricultural use was a late, unstyled D that was shipped in August 1933. The 1935 Model D was the first tractor in commercial production to offer a three-speed transmission, as well as several other changes. Late in the same year, engineers solved some issues with the 6-spline rear axles by introducing new 12-spline axles. Changes and options for the Model D wheels kept coming. By the end of the late D unstyled series in 1939, there were options for steel wheels or rubber tires on either end of the tractor; a variety of lugs, scrapers, and extensions for steel-wheeled tractors; a few sets of skeleton rear wheels; reversible rear wheels that narrowed the tread to 47 inches for row crops; and more.

Options on the mid-D included muffler and spark arrester, auxiliary air cleaner, exhaust elbow, drive wheel scraper, extension rim scraper, radiator guard, drawbar shifter for sidehills, electric lighting equipment, citrus grove fenders, and a speed reducer. Options on the late D included electric lighting; Bosch magnetos; a hot manifold routing exhaust to the right side of the tractor; reversible rear wheels for row crops; radiator shutters that could be controlled on-the-go; short front axle for row crops; 28x6 angled flat spoke front steel wheels; 24x5 round spoke front steel wheels; cast disk front steel wheels for orchard service; and round, spoke front wheels for rubber tires.

As early as 1925, a Model D was modified for industrial work with the addition of hard rubber tires, higher speed final drives, and wheel weights. A few Model D tractors were modified into graders between 1928 and 1931. An industrial Model DI was painted highway yellow and built in small numbers between 1936 and 1941. It was followed by row crop versions: the Model AI and Model BI. Sales were slow and all three models were canceled in 1941. According to www.johnnypopper.com, Waterloo released 87 unstyled Model DI tractors. The first was manufactured on December 9, 1936. The DI had turning brakes with a lever handle to lock the brakes, a heavier drawbar, a padded seat, heavier rear end, and different speeds for final drive sprockets.

For collectors of two-cylinder tractors, the John Deere Model D has several highly prized specimens. The Spoker D with a spoked flywheel stands at a near-legendary level. The status comes from the fact that these were the first true John Deere-designed tractors in successful commercial production. Only 50 were manufactured in late 1923, followed by roughly 760 in 1924 and 2,000 in 1925.

Within the Spoker D group, the first 50 had ladder-style sides on the radiator, fabricated front axles, steering wheels with four holes per spoke, and instructions stenciled on the rear of the fuel tank. Changes began with the 1924 production year. A one-piece malleable cast-iron front axle replaced the fabricated axle. The radiator now had flat sides, the steering wheel had three spokes, and some lettering was changed.

The Waterloo factory's assembly line is a flurry of activity in this 1924 photo.

Starting on October 8, 1924, the new Model D for 1925 had a revised main case and a smaller flywheel. The new main case had provisions for the first John Deere power take-off (PTO) as an option. The steering wheel now had two holes and the steering shaft was modified into a two-piece system that provided more space between the shaft and the flywheel. The solid 24-inch flywheel was soon dubbed the "nickel hole." It had a nickel-size hole at each end of the stress slot. It was better than the spoked flywheel, but problems soon developed. The stress slot was short and the relief hole was small. When wedges were driven into the stress slots to assemble the flywheel onto the crankshaft, it was prone to cracking. Deere modified the design and solved the problem in less than a year, but it already had sold approximately 2,400 of the nickel hole D tractors. Most of the nickel hole flywheels were replaced in a special program, which left the few that survived as among the biggest prizes that a Model D collector can find.

A third special member of the early unstyled Model D family is known as the "corn borer" special. It

continued on page 30

This 1930 D is shown working a wheat field with a power binder.

THE (ALMOST) GREAT DAIN

From 1910 to 1923, Deere & Company acquired and integrated several factories that had supplied products to sales branches. The firms included Deere & Mansur, Moline Wagon Company, Davenport Wagon Company, Fort Smith Wagon Company, Van Brunt Manufacturing Company, Marseilles Company, and the Dain Manufacturing Company.

In early 1912, Deere & Company asked Max Sklovsky, head of the engineering department, to evaluate the recent tractors being offered for farming. Soon, Deere engineers began creating their own tractor concepts. Joining Sklovsky in the design efforts were C. H. Melvin, Joseph Dain Sr., Elmer McCormick, and Walter Silver.

The first attempts were variations of the motor-plow or tractor-plow concept. Plows were mounted under the tractor frame midway between the front and rear wheels. On July 1, 1912, Melvin was assigned to design and build one experimental unit. Although space was provided in the Plow Works

An artist's rendition of the three-wheel Dain tractor. Production was authorized in 1917, but was cancelled soon after the Waterloo Boy factory was purchased.

and there was a $6,000 expenditure, there were no other records of the experiment. Evidently, the results were disappointing. Around that same time, Sklovsky produced drawings for a motor plow but the design wasn't taken any further.

A new effort began in June 1914. Deere & Company approved a proposal from Joseph Dain to build an efficient, small tractor that would pull up to three plows. By January 1916, Dain felt one of his three experimental tractors was ready for a serious production test. Dain was sent to Texas for the trial. On March 13, 1916, he wrote from San Antonio, Texas:

"Have followed tractor closely for two weeks. Conditions extremely hard and rough. Absolutely no weakness in construction. Gears, chains, universals, in fact all parts in good condition. Tractor has travelled near five hundred miles under extreme load. Change speed gear a wonder. I recommend to the board that we build ten machines at once."

The board approved. That July, with production in progress, Dain recommended a new motor that was under development in Minneapolis by an engineer named McVicker. The parts were more readily available for the McVicker motor. He also estimated production costs for the tractor would run at $736 to $761, depending on the price of steel. Development work continued and estimated costs increased. Finally, responding to a September 1917 letter, the board authorized Dain to build up to 100 All-Wheel-Drive tractors and to continue experimental improvements. Unfortunately, only a month after the project authorization, Joseph Dain died in Minneapolis on October 31, 1917. His death came shortly after a trip to Huron, South Dakota, for performance evaluations.

Yet, the plan to build the All-Wheel-Drive tractors remained on track. They would be built in the East Moline Marseilles Company factory and Elmer McCormick would be in charge of building them. McCormick had worked closely with Dain on the tractor and engine almost from the start.

Max Sklovsky (seated) and Theo Brown were two of the leading engineers in the development of John Deere two-cylinder tractors. This rare photo is from Jack Kreeger's collection.

However, plans changed in the winter of 1918. After purchasing the Waterloo Boy factory, Deere & Company discontinued the Dain tractor. Theo Brown, who was a promising young superintendent engineer at the Marseilles Factory in 1912, wrote a short history of early tractor development and explained the 1918 decision as:

"1. In 1918, most of the full line implement companies were building tractors, particularly International Harvester Company and Case. It was felt that it was imperative to get into the tractor business at as early a moment as possible in order for Deere & Company to hold onto their plow business. A tractor and plow were usually sold by the same dealer.

"2. That the Dain tractor, while a good tractor, was high priced and also that it would require considerable time to tool up for manufacture.

"3. That Mr. Joseph Dain, Sr., who was responsible for the Dain tractor, had died.

"4. That Deere & Company had purchased the Waterloo Gasoline Engine Company and thus had a factory, a tractor, and an organization that was functioning and so was in the tractor business at one quick stroke.

"5. It should be emphasized that in his thinking about tractors Mr. Dain was ahead of his time, for he insisted that tractors must be made better in every way and that price was not the main objective. Many features of the Dain tractor were better than those of most tractors of that period."[2]

The All-Wheel-Drive tractor, affectionately known as the Dain, could pull a three-bottom moldboard plow in most soil conditions. A four-cylinder gasoline engine delivered 12 horsepower at the drawbar and 24 horsepower at the belt. It had three steel wheels with cleats. Two front wheels steered and pulled, and a single wide rear wheel pushed. Without a differential, power was transferred by two drive chains to the rear wheel and one chain to each front wheel. The transmission provided two speeds, forward and two reverse. The operator could change speed under load without shifting gears, stopping, or depressing the clutch. The operator's seat and steering wheel were offset to the right to give excellent visibility for plowing. A 30-inch-diameter pulley on the left side provided a belt speed of 2,190 feet per minute. A drawbar was suspended beneath the platform. The All-Wheel-Drive tractors were shipped to dealers in the Dakotas. The only complete All-Wheel-Drive tractor today is kept in the John Deere Collections Center in Moline. It was the 79th production unit. A nearly complete All-Wheel-Drive is owned and operated by a Sycamore, Illinois, tractor club.

[2] Deere & Company's Early Tractor Development, by Theo Brown. Two-Cylinder, Grundy Center, IA, 1997

Continued from page 27

was a standard Model D equipped with PTO and stamped with a USDA identification number. In 1927, the USDA purchased 440 Model D tractors and 360 Fordson tractors. They were for part of an equipment rental program that was offered in New York, Pennsylvania, Ohio, Michigan, and Indiana. For $1 an acre, farmers could park the horses and rent a tractor, a plow, and a PTO-driven "stubble beater." It helped farmers combat the growing infestation of European corn borers by reducing or burying crop residue. This special Model D also became a collector's item.

The following year, Deere built 100 experimental Model D tractors and placed an X as a prefix beside the serial number. They became known as the 1928 Exhibit A model. The experimental changes worked very well in field conditions. Deere developed a few more changes and applied them to a second set of 50 experimental tractors in 1930. The serial number for the second round of tractors began with the letter B. Most of the Exhibit B tractors were shipped to Montana or Arizona. Ten of these were equipped with crawler tracks, which was Deere's first venture into that emerging market.

The late unstyled Model D incorporated most changes that had been tested on Exhibit A and Exhibit B units and was introduced in 1931. Sales were excellent at first. More than 5,600 were built for 1931. However, the Great Depression and Dust Bowl conditions nearly forced Deere to cancel tractor production in 1932 and 1933. Records show approximately 320 Model D tractors were built in 1932 and more than 480 were built in 1933.

A Model D, with hard rubber tires, pulls a road scraper in a field demonstration near Lincoln, Nebraska. The photo is dated February 13, 1926.

This 1952 John Deere Model R tractor is owned by Mike English of Carlyle, Illinois. It is a massive, standard-tread, diesel-powered tractor that still earns respect at tractor events. The crankshaft alone weighs 220 pounds. The tractor's complete restoration was a one-year project for English.

The Styled D

Nine years of generally good tractor sales followed the Great Depression years. Waterloo workers averaged a production of 4,100 Model D tractors. On April 7, 1939, Deere began building the first styled D with serial number 143800. Because the tractor was now advertised as the styled D, all preceding Model D tractors became known as unstyled units. Deere built 46,900 styled Model D tractors in the next 15 years and shipped them to most areas of the United States and Canada. They were well suited to the dry areas of western Canada, where many still survive.

Wartime demands diverted the supply of steel and other supplies for awhile. Styled Model D production plunged from 2,990 in 1941 to 2,571 in 1942 to only 557 in 1943. Supplies and production increased in 1944 to

more than 3,400. Production hit a postwar peak in 1948 at more than 8,700 tractors. As a hard-to-find symbol of an era, the 1943 D is a highly collectable tractor today.

Collectors also search for end of production models, as well as the first models in a new line. In the case of the styled D, there were two sets for end of production models. The end of the 30-year Model D production run came on March 25, 1953, when the last styled D that had been built on the assembly line was recorded as built. However, that isn't the end of the story. Later that year, spare parts were used to build 92 more styled Model D tractors that were assembled with hand tools to meet a few last orders. The assembly point was a street location in Waterloo between a truck shop and a milling room. Most Streeter D tractors were exported to Cuba, about a third were used in the rice fields of Arkansas,

There are belts galore in this 1924 photo from inside the Waterloo factory.

and a handful went to Kansas, South Dakota, and Saskatchewan. A few of these last styled D tractors remained in stock in Waterloo for many years and could be shipped when a customer or dealer insisted on one more Model D tractor.

Model R

In June 1948, Deere & Company introduced the mighty Model R tractor to dealers in Winnipeg, Canada. The postwar economy was booming and farms had a taste for power and technology. Diesel engine technology had come a long way since Rudolph Diesel's first compression-ignition, oil-burning engine in 1892.

IHC and Caterpillar started working on a diesel-powered engine during the late 1920s. Caterpillar had a diesel tractor in 1932. IHC delivered the T-40 diesel crawler in 1934. International's next product, the 1935 WD-40 standard tractor, was a direct competitor to the John Deere Model D. Within three years, Waterloo had built two experimental diesel tractors. In early 1940, Deere & Company ordered a design for a completely new tractor with a diesel engine. It was to be larger, more powerful, and more versatile than Model D. Prototypes,

known as the MX tractors, were tested in 1941, 1944, and 1947. In August 1947, final changes were specified and the new diesel was declared ready for production.

On January 12, 1949, the first production Model R tractor, serial number 1000, was shipped from Waterloo to the Deere & Company experimental department. In March of that year, it was picked up at the factory by Louis Toavs of Wolf Point, Montana. Toavs traded it in 1955 for a new Model 80, but he acquired the tractor again in 1980. Today it is part of the huge Toavs collection of John Deere tractors.

Historically, it was a quiet week when the first Model R was shipped. Farm radios were probably tuned in to hear host Bing Crosby on Philco Radio Time on that night in January. His guests were Johnny Mercer and Peggy Lee. A handful may have picked up the television premiere of Arthur Godfrey and His Friends on CBS. In Washington, J. Edgar Hoover, director of the FBI, was getting ready to meet Shirley Temple on January 20, 1949. His gift to her was a tear gas-filled fountain pen.

The Model R was more than a match for the competition. It was the most massive tractor that Deere & Company had built. Almost every part was rugged. It

An over-the-hood view from the driver's seat of the mighty Model 80 diesel. The small red cap in the middle of the hood is the gas tank cap for the V-4 pony motor.

Ready to rumble with dual rear wheels, this is a 1956 Model 80 diesel that was fully restored by owner John Jensen of Lakewood, New York. Jensen purchased the tractor in original condition from a Minnesota farmer in 1990. The tractor has been completely rebuilt. It was set up for dual wheels to provide extra traction with heavy plowing.

had more power than the Model D, and fuel economy was terrific. It used a 416-cubic-inch, two-cylinder diesel engine that had the same bore, stroke, and intake valve dimensions as the D-8 Caterpillar. To start that huge engine, engineers also mounted a gasoline-driven, high quality, two-cylinder pup engine. The little high-performance engine with 23 cubic inches built up about 10 horsepower at a speed of 4,000 rpm. While it warmed up, the starting engine was warming coolant shared with the diesel engine and the exhaust heat was pre-warming intake air for the diesel engine. Once the gas engine was warm, the operator slowly pulled a lever to engage the diesel engine, which also engaged the decompression lever. After the diesel engine coughed and turned over a few times, the decompression lever could be released while the diesel was given a little more throttle. Once the diesel engine warmed up, the operator had to remember to shut off the starter engine. The diesel might take ten minutes to warm up, but after that, it would out-pull anything in the class. In cold weather, the start-up procedure might take a half hour.

The Model R went through the Nebraska tractor test station during April 1949. It shattered the fuel economy record that Caterpillar had held. For power, it produced a corrected maximum of 50.96 horsepower at

the belt and 45.69 horsepower at the drawbar. Those ratings surpassed the IHC WD-9 tractor by several horsepower while doing it with less fuel. It was rated as 28 percent better fuel economy than the WD-9.

Deere & Company gave the market a variety of choices with the R. It had a wide range of tire and wheel options. Besides working the Great Plains, the Model R could be set up for rice and cane fields. Even steel wheels could be supplied for very rough conditions and for deep rice fields. Hydraulics and PTO were optional. A radiator shutter, hour meter, and wheel weights were other options. Deere even offered a special heavy load or industrial transmission with reduced third-, fourth-, and fifth-speed gears.

The Model R contained several firsts. It was the first diesel tractor offered by John Deere. It also was the first John Deere with a live PTO and a form of live

Stay clear when the wheels start to roll! These wheels get a workout at a plow day.

hydraulics. The hydraulic pump was driven by the PTO and controlled by a second clutch. It also offered the first all-steel cab for an operator as an option.

This powerhouse had a listed weight of 7,400 pounds, which was over 2,000 pounds heavier than the Model D. It was 17 inches longer, 13 inches wider, 17 inches taller, and it had a five-speed gearbox compared to the four-speed in the late styled Model D. Owners discovered it would pull more plow bottoms than a D in nearly all conditions.

The 1949 production run of Model R production amounted to 1,415 tractors. In late 1951, two were built as an industrial version with yellow paint, according to the website, www.johnnypopper.com. At least one Model RI still exists. Most Model R tractors were sold to farmers in the wheat and small grain regions of the High Plains and Prairies. The production total shipped to the United States and Canada was 17,563 models. Another 3,570 tractors were exported. Sales of the Model R peaked at 5,800 in 1952 and ended on August 26, 1954, with 3,200 units. By then, Deere & Company was satisfied that it had a successful market entry with the big standard-tread diesel. As engineers readied a set of upgrades that would

be introduced as the Model 80, the company phased out its original standard tread, kerosene-fueled Model D.

Model 80

On June 27, 1955, *Time* magazine featured Dag Hammarskjold on the cover. A devastating tornado roared down the North Platte river valley in western Nebraska on this same day. In Waterloo, Iowa, the first John Deere Model 80 came off the assembly line. It was close to the end of the two-cylinder era. Engineering, tooling, and technology had come of age in farm machinery. The postwar economy was booming. Veterans were putting rubber to the ground and steel to the furrow. Bigger was better. Farmers wanted more power, more options, and superior implements for larger fields and larger farms. Deere & Company factories rolled out what the economy wanted—tractors with more horsepower than ever before. More than that, it offered a quick progression of new models from the very powerful, large diesel tractor to small, high-performance tractors for intensely productive small farms. This quest for power created a new kind of war. Man and machine would make the earth produce more food than ever before. In another five years, the two-cylinder era would come to a close. Progress had determined that more cylinders were better.

The Model 80 holds the distinction of having the shortest production run of any John Deere two-cylinder tractor. It came out just before the 1955 mid-summer shutdown. Waterloo built about 3,500 in the next 54 weeks. On July 11, 1956, the run was finished. The Model 80 is the rarest of the large, standard-tread Deere-built tractors.

The Model 80 had many outward similarities to the Model R, but at a distance the teardrop-shaped flywheel cover and John Deere medallion embedded in the hood above the grille were distinctive. Impressive as the 51-horsepower Model R had been, the Model 80 was bigger and better. Today, the five-ton Model 80 is a collectable heavyweight and requires a big trailer and large sheds between showing events. In its day, it set records and had many new features. In later years, it developed an excellent history as a powerful, reliable workhorse with amazing fuel economy. It out-powered the unstoppable Model R by about a third. At the Nebraska tractor test facility, the 11,485-pound Model 80 set records in October 1955 with 67.65 belt horsepower, 61.67 drawbar horsepower, and a maximum pull of 7,394 pounds. On a long, sunny day in good conditions, a farmer with a Model 80 could cultivate 125 acres of land and raise a few eyebrows along the fence line.

Model R operators had a six-speed transmission for a faster pace on the improving roads. Under the hood, engineers had bored out the two cylinders another 3/8th of an inch. The Model 80 also had a significantly stronger three-bearing crankshaft with a center support. The Model R's little two-cylinder starting engine was replaced with a beefier four-cylinder engine known as the V-4.

The former Powr-Trol hydraulic system in the Model R had a major upgrade and was renamed the Custom Powr-Trol in the 1957 models. The Model 80 was the first Deere tractor that could be ordered with a dual hydraulic box to give separate hydraulics for each cylinder and the ability to separately power two hydraulic circuits. In fact, buyers could order the tractor without a PTO and have the hydraulics, or they could order it with the PTO but without the hydraulics. The dash had a fuel gauge and a new speed-hour meter.

The base price for a Model 80 was $4,205. Buyers could customize it with a big choice of options including power steering, live power shaft, dual controls and breakaway couplings, remote cylinder, creeper gear, wheel weights, and a cigarette lighter. Most were delivered as fully loaded. Dealerships also had several options that could be installed including an enclosed steel cab, a muffler extension, and warning lamps for the road.

Model 820

The first Model 820 was built on July 10, 1956. That same day, the news reported 650,000 U.S. steel workers had gone on strike and the U.S. had conducted an atmospheric nuclear test at Bikini Island. In sports, the National League won a 7-3 All-Star Baseball game at Griffith Stadium in Washington, D.C.

For tractor buyers, Custom Powr-Trol was big news. The 20 Series had an improved version of dual live hydraulics that dated back to the Model R. Custom Powr-Trol was completely independent of the transmission

This early 1957 John Deere Model 820 Rice Special is owned by S. D. Palmer of LaFayette, New York. The diesel engine was completely rebuilt before the restoration work could be completed around 2003. The four-cylinder pony engine and transmission were still in good condition.

This rear view shows the unusual fenders and oversize 26-inch rear wheels for heavy work in flooded rice paddies.

After July 1957, the tractor's entire dash was black. The earlier tractors became known as the Green Dash Model 820.

clutch. It included Draft Control, a feature that forwarded implement loads through the top link (sometimes called the free link). It actually controlled an implement's working depth by pushing on the hydraulic control valve. These tractors also had PTO, which enabled an operator to have two rear hydraulic outlets with an emergency disconnect system.

The Powr-Trol-equipped Model 820 weighed 8,150 pounds and looked like an upscale twin to the Model 80. The 820 had a new touch of style—bright yellow hood side panels. It also had the elephant ear fenders common to Waterloo's row-crop tractors.

The Model 820 had the same engine as the Model 80, but it had stronger connecting rods. The power steering system was new. There were minor improvements in cooling and lubrication, as well as larger 10-inch brakes. An optional "Float Ride" seat could be adjusted for the operator's weight. The steel cab had better sealing and soundproofing to reduce the noise inside the cab. Other options included an air intake pre-cleaner, an oval muffler, a foot-operated 'gas' pedal/accelerator, a cigarette lighter, and a creeper first gear. In creeper gear, the 820 could pull a combine at 1 3/4 miles per hour.

The Model 820 Rice Special became available as a package deal. The buyer received either 15-34 or 23.1-26 rear tires with cane and rice treads. The front tires sported a single rib. The rear axle and brakes were protected from

mud by special shields. A new decal declared the unit was a Rice Special.

A few Model 820s were equipped for industrial work, but there wasn't a specially designated version. The industrial 820 had yellow paint, larger rear tires, heavy-duty front axle, heavy-duty drawbar supports, drawbar extension, foot throttle, and weighed an extra 1,600 pounds.

A power upgrade for the Model 820 became necessary when the Waterloo factory introduced its 20 Series Model 720 diesel shortly after the Model 820 hit the market. A power boost had given the Model 720 a nearly identical power output. In Waterloo, engineers tweaked and boosted power for the Model 820 by 12 percent to reach 75 horsepower. They modified the pistons for more compression, improved the engine's breathing capacity, increased the size of the fuel line, and improved the fuel injectors. Other changes made the tractor easier to start and more reliable. On the early 820, the dash was green. Waterloo built 3,100 tractors with a green dash. The next 3,900 became known as the Black Dash 820. Production for this model ended on July 23, 1958.

Model 830

The Model 830 became the kingpin of the entire John Deere two-cylinder tractor line. It was sometimes known as Mr. Mighty or Big Daddy. The birthday of the Mr. Mighty series was August 4, 1958. On the radio that day, "Hard Headed Woman," a song by Elvis Presley, was the chart topper. In fact, the first Billboard Hot 100 music list was published that week. In the far north, America's atomic submarine, the USS *Nautilus*, was on its maiden voyage under the North Pole ice.

The Model 830 with Powr-Trol weighed 8,150 pounds, which was the same as the Model 820. However, these were the green-and-yellow farm equivalent to the late 1950s Cadillac—big, quiet, and comfortable. In fact, the entire 30 Series, including the 830, had a new oval muffler that made quieter tractors.

Upgrades to the Model 830 and the 30 Series were for comfort, convenience, and productivity rather than power. The PTO, rated for 75.6 horsepower, could handle any load a farm would need. Operators had a more comfortable seat, better controls, and improved instrumentation. They had a foot-pedal accelerator in addition to the usual hand-operated lever accelerator.

Where winters were mild, many buyers chose to install the new battery-operated 24-volt electric starter option on the 830. Where winters were cold, the gas-fired V-4 starting engine was a better choice than a battery and electric starter. The V-4 even pre-warmed the big diesel engine by sharing the coolant to help the diesel get started more easily in cold weather.

Production included 127 industrial versions of the 830, according to www.johnnypopper.com[5]. There were 82 with electric start and 45 with the V-4 starting engine. They were painted yellow and upgraded to heavy duty in the axles, front end support, front wheels, and drawbar. The most spectacular tractor for its time was the 830 industrial modified with a side-mounted operator station so that it could mount and operate a Hancock self-loading scraper.

On July 14, 1960, the era of the big, standard tread John Deere tractor that had started with the 1923 Model D was over. Approximately 6,900 Model 830 tractors had been manufactured. These last were the largest, most powerful two-cylinder tractors that Deere ever produced. By then, only one other two-cylinder assembly line was still operating at the Waterloo factory. On September 27 of that year, the Waterloo factory closed forever.

[5] www.johnnypopper.com/weirddeere/The_830.html

This Model 830 diesel was built in 1958 and is owned by Bud Reifsneider of Royersford, Pennsylvania.

This "hybrid" sure can pull! The engine in this fully restored Model 830 was replaced in 2007 with an eight-cylinder Challenger 3208 engine that produces about three times the horsepower. Owner and collector Elmer Friesen of Rosenort, Manitoba, plans to accept challenges anytime and anywhere.

CHAPTER 3

WATERLOO FACTORY
BIG MAMA TRACTORS
MODELS A, 60, 620, 630

The Waterloo Row Crop Tractors

After the successful launch of the standard-tread Model D tractor, the Deere & Company design engineers soon started a second tractor concept. This tractor featured a non-standard tread (wheel spacing) to better suit the needs of row crop production. Between 1928 and 1937, the engineering department developed three fundamental sizes of row crop tractors for production at the Waterloo factory. The spacing of the front and rear wheels on a row crop tractor could be equal, but that was the exception. Some tractors had single front wheels, some had a tightly spaced pair of front wheels, and most had much smaller front wheels. Often wheel spacing was adjustable to suit the spacing of rows.

Deere's first row crop tractor family started with the General Purpose tractor that was introduced in 1928, about a year before the stock market crashed (see Chapter 4). The models in this chapter are members of the mid-size row crop tractor family that was introduced with the Model A in 1934 during the middle of the row crop development era. After the economy began to recover in 1937, Deere introduced a third family of larger row crop tractors (Chapter 5). Finally, in 1939, a family of tractors suited for the smallest commercial farms was launched (Chapter 5). By the end of the era in 1960, the Waterloo factory had built just over 1 million two-cylinder tractors.

For perspective, the standard-tread John Deere Model D had tremendous power for plowing from sunrise to sunset in broad open fields, but there were other needs. Even more farms required a general purpose tractor for a range of applications. Small farms wanted small tractors. Other farms wanted tractors for corn, cotton, vegetables, orchards, or other row crops. Tractor manufacturers responded. By 1939, the Waterloo factory was building five separate streams of tractors. Pre-war production at Waterloo peaked with the five assembly lines building approximately 55,000 tractors in 1941, which averaged more than 1,000 tractors each week.

Along with adaptable wheel spacing, the Deere & Company row crop tractors burned different fuel than the standard-tread Model D or its successor, the Model R. Most row crop tractors burned gasoline or had a two-stage fuel system that burned gasoline followed by distillate. Kerosene, the standard fuel for Model D tractors, was mostly derived from coal. Distillate, also called tractor fuel, came along a bit later. It was similar to kerosene, but it was derived from petroleum and it was cheaper than gasoline. Since distillate had a low octane, it wouldn't start a tractor

OPPOSITE: This rare tractor is one of the first 300 John Deere Model A row crop tractors built in 1934 and has the four-bolt pedestal. It is owned by Steve Slack of Swansea, Illinois. It was purchased after a rollover and was completely restored by Steve Slack and Ernie Restoff in the late 1990s.

but it was okay after the engine was hot. Engineers came up with a workable and more economical fuel system by putting two fuel tanks and a water tank on the row crop tractor. These "All Fuel" tractors started with gasoline from a small tank. Once the engine warmed up, operators switched a valve to the main tank of distillate. The tractor didn't develop as much power on distillate, but it was much cheaper to run. The water was used to reduce detonation or pre-ignition. After the two-cylinder era, when gasoline became as cheap as distillate, the All-Fuel engines were discontinued and engine compression was increased to get the full horsepower potential from gasoline.

A few two-cylinder row crop tractors ran on liquid propane (LP) gas, which was high octane. It burned hotter, generated more power, and was available at an economical price in some southern U.S. farming areas. Approximately 27,100 LP row crop tractors were built, which is about 2.7 percent of the total Waterloo row crop tractor production. The remaining LP tractors are harder to find and more valuable than the gas and All-Fuel models.

Row Crop Variations

As Deere & Company met the needs of the market, the basic two-cylinder row crop tractor went through many variations, accessories, and options. This chapter heading identifies only four models. In fact, there were at least 61 versions in the mid-size row crop tractor family. They are classed by model, style, fuel, and application. Similar numbers of versions apply to the families described in Chapters 4, 5, and 6.

The following is a quick guide to the basic configurations for wheel tread and height.

The 'ordinary' row crop tractor rode on three points, which are referred to as the tricycle-type shape, with the front carrying two closely-spaced wheels. This configuration had a basic designation, such as Model A. After the model letter, a second or third letter could be added to mark a significant variation. Thus:

- Some had a narrow, single front wheel under the nose. This was designated as the N version, such as Model AN or Model 60N.

- Some had an adjustable-width front axle. The tread of the front actually could be wider than the rear. This was designated with a W.
- Some were equipped to be tall for more clearance. These were Hi-Crop tractors, designated as HC.
- Some HC tractors had a single nose wheel. These narrow-nosed, high clearance tractors are designated as NH.
- Some were built as both tall and width-adjustable on the front axle. These became the WH tractors.
- Some were standard tread with a solid beam axle under the front pedestal and fixed wheel spacing. They used a row crop chassis, components, and fuel, and could be a good option to the larger Model D. This was designated as R.
- Some standard-tread row crop tractors were equipped to work in orchards. They had less ground clearance and were streamlined to reduce snagging on branches. They were designated as O or OS.

Model A

A serial number plate was fixed to the first John Deere Model A tractor on April 8, 1934. The Great Dust Bowl storm began blowing a month later on May 13. There was a bit of cheerful news—movie theaters introduced a new character, Donald Duck.

Despite a launch in desperate times, John Deere's mid-size row crop tractor became a huge success. There was a large number of models built and many still survive; therefore, the Model A family is among the least valued John Deere two-cylinder tractors. Deere & Company only manufactured about 2,800 Model A tractors for 1934. However, production rallied quickly once America was coming out of the Great Depression. Waterloo built more than 65,000 tractors in the first five years. The styled Model A was introduced for 1939. Waterloo built nearly 100,000 of these early styled Model A tractors through 1947. There were 90,000 late styled Model A tractors built up to the end of the series in 1953. The family continued on until 1960. Over 26 years of production, Waterloo built about 425,000 tractors in

This detail shot shows the positioning of the John Deere two-bottom, 14-inch trip plow for plowing two furrows. Rubber belting protects the plow wheels. The tractor was manufactured with 50-inch steel skeleton wheels that were replaced at some point with these replica French & Hecht rims and rubber tires.

the mid-size family. They were coming off the Waterloo assembly lines at rates of about 100 per day.

Development of the Model A family began in early 1931. It was determined that the new tractor should be able to handle crops spaced in rows 20 to 42 inches apart. It also needed to have high crop clearance and be more adaptable than the disappointing General Purpose Wide-Tread (GPWT). Other changes engineered into the Model A included the first use of a hydraulic cylinder to provide a power lift, over-the-hood steering, and adjustable tread width for the rear wheels.

Waterloo tested ten pre-production prototypes of the unstyled Model A in the spring of 1933 and designated them as AA tractors. The AA tractors were rebuilt into regular production tractors and issued new serial numbers. After a field demonstration on June 14, 1933, company officials decided the four-speed Model A version was a winner and cancelled plans for a three-speed

version. Production was slated to begin in December 1933 at a rate of 50 per day, but delays set back the start of full production until the first week of April in 1934.

The new 1934 Model A was the first tractor in the two-plow class to offer adjustable rear wheel tread from 56 to 80 inches. The narrow tread virtually eliminated side-draft when it pulled two 14-inch plows. At 80 inches, it was suitable for working in two rows of corn or cotton.

Optional hydraulic Power Lift probably was the most important of many innovations. It was exclusive to John Deere and it eliminated the back-breaking labor of raising and lowering implements. The one-way hydraulic piston lifted up and cushioned the drop of the implement, but it couldn't exert downward pressure. Research showed it enabled a farmer to increase production by about 10 percent compared to hand-lifted implements. According to Ken Burns, a Nebraska collector and two-cylinder historian, most owners of the early row

This 1937 Model A on tiptoe lug steel wheels, as seen from the rear side corner, is owned by Roger and David Dills of Brockport, New York. This tractor worked in the North Carolina fields before it went to upstate New York for restoration in 2001. It's a ground-up restoration by the father and son team.

crop tractors called this the rockshaft accessory rather than Power Lift.

Initially, the Model A was an upgraded General Purpose Wide-Tread (GPWT) tractor that Deere & Company began building before the Depression hit. On the assembly line, Waterloo even allowed 1,100 of the first Model A tractors to slip through with a mixed designation. The Model A was in effect a General Purpose tractor with a two-wheel, tricycle-style front axle. In the two-plow class, it was the first tractor with an adjustable tread width for its rear wheels. Rear wheel spacing could be 56 to 80 inches wide to give farmers flexibility to work in 40- or 42-inch rows of corn or cotton.

The open fan shaft was a weak point. It was prone to breaking because there was nothing to support the front of the fan shaft. When it became unbalanced or if the clutch wore, the housing would break. In mid-1935, Waterloo began building an enclosed fan shaft to fix the problem.

There are few open fan shaft models that survive today. They are easily recognized and prized by collectors.

Customers and dealers soon asked for variations of the Model A. Vegetable growers with narrow rows wanted a version with a single front wheel. Growers with crops in raised beds wanted a version with an adjustable width front axle. By mid-1935, both options were available.

A big change in late 1937 made it much easier to change the tread. Waterloo had been manufacturing axles with 10 splines to hold the wheel. They increased the number of splines to 12 and adopted a new method of holding the rear wheels by casting a tapered hub into the wheel. Changing the tread was faster and there was greater clamping force after the tread was adjusted. Around 1942, Waterloo increased the pattern to 15 splines.

Wheel variations for the Model A were perhaps the largest ever offered. The Model A could have steel wheels front and rear, rubber tires front and rear, or any

Here is a close-up of the cast lugs on steel rims. Each lug weighs about 20 pounds. Note the vertical steering wheel and single square tube support for the driver's seat. This is a very basic operating system.

This is a view of the simple controls and clean lines from the operator's platform on the 1937 Model A. It has only two gauges, located near the driver's right foot, for the oil pressure and water temperature. There are two fuel tanks. The smaller one, closest to the operator, is for gas. The larger tank, located over the engine, is for distillate.

combination the operator requested. The wheel could be supported by spokes, by cast, or by a solid disk. The steel wheels could be flat, tiptoe, or skeleton, depending on the traction requirements. Some collectors have specialized in variations of wheels and splines on Model A tractors.

Early in 1938, higher clearance versions of the AN and AW were offered for California vegetable growers. These axles had taller wheels and taller axle knees on the front. A few special units were equipped with special rear axles and special offset wheels that enabled them to hold a tread of 104 inches.

The standard-tread AR became a popular variation during this period. The first AR was built on April 25, 1935. After 17,000 were built, it was upgraded to a styled AR as of December 10, 1940. Waterloo built another 10,400 of these, including 500 with gasoline two-cylinder engines. The AR stayed in production into November 1948.

Between 1936 and 1941, Waterloo built an industrial version, Model AI, which was based on the AR. According to www.johnnypopper.com, 91 models were built and 68 were released. The other 23 were scrapped or rebuilt into a Model AR. The AI was yellow, more comfortable, and had many modifications. For instance, the front axle was moved back seven inches for equipment clearance and more maneuverability.

The first styled A rolled off the assembly line on August 1, 1938, as a 1939 model. Under the new styling, it was mechanically identical to the 1938 model. Mechanical changes came with the 1940 models. Piston

A John Deere Model AI pulls a railcar at the Deere plant in 1936. The industrial version of the Model A had special mounting pads on the frame for industrial equipment, such as sweepers and loaders.

stroke increased from 6.50 to 6.75 inches. Engine breathing was improved. It gained a new carburetor. It was built stronger with changes to the main case, transmission, and final drives. It now was a close competitor to the IHC Farmall M and a better choice for tough soil conditions. For 1941, the Model A was equipped with a new six-speed transmission.

During World War II, product development for the styled A slowed, but there were some changes from year to year. The 1941 Model A could have a new six-speed transmission if it was equipped with rubber tires. Fourth gear was considered the top speed that was safe for a tractor with steel wheels. The 1941 Model A also had a new Wico C magneto and a new carburetor, the DLTX-53. The changes were successful and Deere built more than 12,000 of these tractors in 1941.

Production was suspended from November 1942 through March 1943 due to wartime shortages. In 1945,

The Model A tractor with a A492 mounted cultivator was ideal for cultivating corn in 1937. This tractor was photographed near Wellington, Illinois, and seems to have a revised seat.

the new Powr-Trol hydraulic system was an important new development. It was the first hydraulic system for farm tractors that could power a hydraulic cylinder and provide precise positioning of the tractor rockshaft. In this early styled period, options included electric start, electric lighting, overdrive, underdrive, a range of lugs and guide bands for steel wheels, extension rims, wheel scrapers, and road bands.[6]

[6] The John Deere Styled 'Letter' Series, 1939-1952, Second Edition, J. R. Hobbs, Hain Publishing Co., 2002, page.19

This rare 1937 AOS was known as the streamliner because of its distinctive styling, which was introduced before the styled era began in 1939. This prime example was purchased by Patrick Bovat Sr. of Oakville, Connecticut, around 1998. The tractor was fully restored by Bovat.

Late in 1945, Deere offered its first major product development for all three families of row crop tractors in about five years. This was the new Powr-Trol hydraulic system. For the first time, a tractor could power a hydraulic cylinder and provide precise positioning. Powr-Trol initially controlled the rockshaft on the row crop

tractor. The supply was limited for a couple years, but many tractors were retrofitted.

The late A was built from 1947 to 1952 and was a very different mid-size tractor. The basic unit was priced at $2,200, or more than twice the price of the early styled A. Twelve-volt electric lighting and start were standard. Batteries for the new 12-volt system were placed in a box above the rockshaft housing and supported the operator's new seat. The starter was relocated into a sealed, weather-safe compartment at the bottom of the main case. Now the flywheel could be enclosed for more safety and reliability. The foam-cushion seat and backrest were much more comfortable than the earlier pressed-steel seat. At the front, Deere & Company introduced Roll-O-Matic, a knee-action suspension system for tricycle tractors that was invented by engineer Harold Sexton. Powr-Trol was a $149 option and the demand was overwhelming. It wasn't long before hydraulic power lift was standard on row crop tractors.

Model A row crop tractors were sent through the Nebraska tests in 1934, 1939, and 1947. For the 1934 test, the Model A generated 24.7 horsepower on the belt and 18.7 horsepower on the drawbar. It had a maximum pull of 2,923 pounds. The tested tractor was on steel wheels and burned distillate. The shipping weight was 4,059 pounds.

A styled Model A tractor is shown pulling and powering a Dain hay press baler in this 1939 photo. The baler still required two men riding on the machine to insert the wires.

In 1947, the Model A went through the Nebraska test again. The distillate tank was gone; the tractor burned only gas. The engine displacement had a slight increase, but power was up again: 38 horsepower on the belt and 34 horsepower on the drawbar. It weighed 6,574 pounds, but the maximum pull was about the same as the 1939 model. The last AR was tested during the same year and it weighed a hefty 7,367 pounds.

Five years later, the heavier styled A had a small increase in engine displacement and a big increase in power. It had 30 horsepower on the belt and 26 horsepower on the drawbar, and its weight was 6,410 pounds. The pulling power had increased to 4,110 pounds on this rubber tire tractor. The standard tread AR version was slightly more powerful and a bit heavier than the styled A. On May 12, 1952, the last Model A gas-powered tricycle tractor rolled off the assembly line. With 118,000

units built in five years, it was the best-selling series and stood with the best in class.

Model 60

The build date for the first mid-size, two-cylinder John Deere Model 60 row crop tractor was March 12, 1952. It was a presidential day. In Hollywood, actor Ronald Reagan had been married for four days to Nancy Davis. In the White House, President Harry S. Truman had renovations underway in four rooms. It was also the day after stunning results in the New Hampshire primary. Humbled by the results and convinced of a higher calling, General Dwight Eisenhower announced his candidacy for President of the United States.

The Waterloo factory built more than 63,000 Model 60 tractors between 1952 and 1956. It was a popular successor to Deere's very popular late styled A and the

The John Deere Model AR is a durable workhorse that has kept chugging along for six decades on some farms. This fully restored 1950 styled model was purchased in its original condition in 2006 by Mike English of Carlyle, Illinois. English has restored it to factory original condition.

It takes an extreme amount of patience to restore a tractor to factory original condition. Mike English did a fabulous restoration job on this 1950 Model AR.

first of five Number Series tractor models introduced by Deere & Company in its transition from Letter Series tractors. About 80 percent were row crop tractors with gasoline engines. About 6,500 had LP gas engines and 5,500 had All Fuel engines. The Model 60 was built with five wheel configuration options.

Compared to the late styled A, the Model 60 offered a small increase in horsepower and good improvements in other features. The dual carburetion, Quick-Change wheel tread, and Roll-O-Matic front end were notable selling points. Operators could also turn the hydraulic system on or off while the engine was running to save fuel and wear on the engine.

True to form, Waterloo obligingly built many versions of the Model 60. Often the various versions were in response to direct requests from buyers or dealers for specific traits. By late 1952, buyers could get a transmission-driven PTO or a new live PTO for $135 and a solid front wheel assembly or a Roll-O-Matic front wheel assembly for $55. The All-Fuel versions offered dual carburetion in January 1953.

In 1954, Waterloo began rolling out the No. 800 series hitch for implements on the Model 60 tractors. It

wasn't up to the standard of the patented Ferguson three-point-hitch, but it was pretty close. The hitch was upgraded again in 1956 to the 801 Traction-Trol hitch. This hitch transferred weight to the rear wheels for integral attachments such as plows.

As with the Model A, Model 60 buyers could choose a standard one-piece front pedestal or an optional two-piece pedestal. The two-piece pedestal enabled a variety of front ends to be fitted, including a 38-inch fixed tread front axle. The two-piece option proved most popular and became standard on 1954 row crop models, while the one-piece pedestal was phased out. Late in 1954, Deere began offering factory-installed power steering for $125 on the Model 60 row crop and Hi-Crop tractors. The demand was overwhelming. According to engineer Gerald Mortenson, Waterloo could hardly keep up with the demand for power steering from day one. Planners had expected a small volume of requests, but quickly upgraded to include power steering on about 40 percent of the new Model 60 tractors. The planning was for a small volume, but it immediately went to 40 percent. The change for growers was huge. For the first time, growers had "one finger" steering available for virtually any field condition.

continued on page 50

THE DREYFUSS TREATMENT

As North America and Europe came out of the Great Depression, it was time to put some style and grace on the two-cylinder workhorses from Deere & Company. Starting in 1939 and over the next three years, all Deere & Company tractor assembly lines changed over to build styled tractors. It became a watershed in the two-cylinder era that separated styled and unstyled tractors for buyers and collectors. Like automobiles, trucks, trains, and planes, the farm tractor had matured. Buyers were more discerning. They wanted form and comfort as well as functionality and reliability.

In 1937, the Model A was enjoying the finest sales year it would ever have, but Waterloo engineers knew the writing was on the wall as they looked at newly styled automobiles. Style soon would be required in the tractor package. Oliver, Massey Harris and Case had introduced tractors with styled sheet metal; the thought was that better-looking tractors would produce more sales. Engineer Elmer McCormick pressed the issue with Deere's vice president, Charles Stone, and in August he received permission to visit New York designer Henry Dreyfuss. Historian J. R. Hobbs writes of that meeting: "Upon

Farm families gathered around the newly styled John Deere tractors at state and county fairs in 1939. The Dreyfuss treatment was an instant success with all ages.

arrival in New York, McCormick immediately contacted Dreyfuss and stated what he wanted. Henry Dreyfuss had never heard of Waterloo, Iowa, or John Deere, for that matter. Yet, the idea of adding style to the utilitarian farm tractor intrigued him. The next day, Dreyfuss accompanied McCormick to Waterloo to see what could be done."[3]

Dreyfuss was born in Brooklyn in 1904 and grew up to become a celebrity industrial designer. Dreyfuss apprenticed with a theatrical designer and opened his own office in 1929 for theatrical and industrial design. The firm, Henry Dreyfuss Associates, still has major corporate clients today. Dreyfuss dramatically improved the look, feel, and usability of major products, from the telephone and vacuum cleaner to locomotives. His approach was described as a mixture of common sense and good science that contributed to the fields of ergonomics and anthropometrics, as well as optimized products to fit well with the human anatomy.

In 1929, Dreyfuss won a phone of the future competition by Bell Laboratories. The result of his collaboration with Bell was the 300 tabletop telephone, which featured a receiver and transmitter in a combined handset that rested in a horizontal cradle. Molded in black phenolic plastic, it was introduced in 1937 and produced until 1950.

In 1933, Dreyfuss designed a new flat-top deluxe refrigerator that was introduced by General Electric. His new style of refrigerator eliminated the previously exposed refrigeration unit by placing it beneath the cabinet. He also designed a new Toperator washing machine for Sears & Roebuck. In 1934, he was featured in a 1934 article in *Fortune* magazine, which had a dramatic impact on the new field of Industrial Design. He was hired by the Hoover Company in 1934 and designed its 1936 Model 150 upright vacuum cleaner with the first plastic hood in Bakelite. Dreyfuss designed a bottle for The American Thermos Bottle Company that appeared in 1936. For Westclox, he designed an alarm clock introduced in 1935 and the Big Ben alarm clock that was introduced in 1939.

In 1936, Dreyfuss' design of a Mercury locomotive debuted. It featured cutout holes in the white-walled driver wheels that were lit by concealed spotlights at night. Two years later, New York Central introduced 10 new streamliner steam engines and cars designed by Dreyfuss for its New York-Chicago run. The New York Central locomotives were an upgraded version of Dreyfuss' Mercury design, but they also featured finned bullet-noses. [4]

After the Waterloo visit, Dreyfuss became involved in the overall design of the Model A. He approached this and future projects for Deere & Company tractors with a five-point philosophy.

1. Utility and safety of the object
2. Ease of maintenance
3. Cost of production
4. Sales appeal
5. Appearance

Dreyfuss worked with Waterloo engineering staff to make the new tractors safer, more comfortable, and easier to operate.

The first new styled A on the Waterloo production line was built on August 1, 1938, as a 1939 model. The first styled Model B was introduced around the same time. Once the Dreyfuss styling theme for Models A and B was completed, it was scaled up or down for all the John Deere tractors from Waterloo and Moline, and for the first M series tractors from Dubuque in 1947. The Dreyfuss treatment was applied to the bread-and-butter row crop tractors between 1938 and 1942, but not to the standard-tread AR or BR. Dreyfuss-designed sheet metal was a blend of form and function. A front-to-rear taper on the hood greatly increased visibility. The end cap of the hood became a perfect place to put gauges. Inline stacks for the exhaust and intake cleared the operator's line of sight.

Dreyfuss died in 1972, but the association between Dreyfuss Associates and Deere remains and continues into the twenty-first century.

[3] The John Deere Styled Letter Series, pub. 2002, by JR Hobbs, @ Green Magazine, page 8

[4] Henry Dreyfuss, FIDSA, 1904-1972 , www.idsa.org/webmodules/articles/anmviewer.asp?a=247

This 1954 John Deere Model 60 is owned by Bernard Molleur of Beacon Falls, Connecticut. It was totally restored around 1998. Since then, it has collected a roomful of trophies and banners from antique tractor pulls.

Continued from page 47

Sheet metal for the hood and fuel tank on the Model 60 changed three times over the production run. The earliest hood was like its predecessor on the Model A. The late hood was similar to the Model 620 that followed.

The first 1,900 units of the standard-tread Model 60 that replaced the AR and AO were similar to the models they replaced. However, the models changed in December 1954. Engineers raised the driver's seat on both tractors and made a number of other improvements. The high seat Model 60 standard for 1955 and 1956 was equipped with power steering, slightly taller tires, and a seat design similar to the new Model 70 row crop series. The high seat improved the operator's comfort, protection, and field of vision. It had taller tires, greater vertical clearance, and an optional adjustable-tread front axle that combined the best of the standard and row crop features. Most of the standard all-fuel models were shipped overseas. Only 15 each of the low-seat and high-seat versions remained in North America. In the family, the Model 60 Hi-Crop is the rarest. Waterloo only built 212 of these: 135 All-Fuel, 62 gasoline, and 15 LP. They were shipped to Florida, Alabama, and Louisiana.

Model 620

The Model 620 was unveiled as the successor to the Model 60 for 1957. The build date for the first unit was June 7, 1956. On the day that the Model 60 had its debut, Billie Holiday was performing in New York City and Bing Crosby was in Los Angeles. Elvis Presley's "Heartbreak Hotel" was at the top of the music charts. Historically, America had been shaken to its roots two days earlier by a federal court ruling that racial segregation on Montgomery buses was unconstitutional, and on this day President Eisenhower was admitted to the hospital for emergency major surgery to deal with ileitis.

Buyers soon learned that under the hood and paint, the Model 620 was a very different tractor from its predecessor. The surface sheet metal was unchanged from the last edition of Model 60, but this tractor had a distinctively different paint scheme. The Model 620 had the new two-tone, green and yellow paint scheme that was associated with the 20 Series rather than the solid green of the Model 60.

Dealers and buyers discovered that the 620 had about 20 percent more horsepower and consumed less fuel than the Model 60. It could pull equipment a bit faster and it could pull slightly wider equipment than before. It had a much better engine block equipped with aluminum pistons. The cylinder bore had stayed at 5.5 inches and the 6.375-inch stroke was slightly shorter, but the engine speed was cranked up 15 percent to 1,125 rpm. Engineers had moved the spark plugs to the cylinder head and put in a distributor drive pad that enabled it to be driven by the camshaft rather than governor gearing. They had strengthened the whole tractor to cope with the extra power and had eliminated the old petcocks for hand-start.

Along with the paint and power, operators found themselves more comfortable. They could order the Model 620 with the new Float-Ride seat that had two foam cushions and a shock absorber. Hydraulics that were introduced on the Model 80 were incorporated into the Model 620. The Custom Powr-Trol was upgraded and allowed control of three separate hydraulic circuits. These circuits included front and rear rockshafts, two separate remote cylinders, or a new three-point hitch with top-link sensing for load and depth.

The Model 620 was available in configurations familiar to Model 60 owners, such as row crop, standard, and Hi-Crop. Standard tread was a continuation of the high-seat 60 Standard. Most options were available for all three versions. The Model 620 Standard did not have the front rockshaft, but that feature was found on the 620 Orchard.

The 620 Orchard was a bit of an oddball. It had the Model 620 engine and driveline components, live PTO, and all the options, but in other respects it was a continuation of the earlier Model A and 60 Orchard series. Deere continued building the 620 Orchard without an update to the end of the two-cylinder era in 1960. About 720 Orchard series tractors were built with a mixture of gas, All-Fuel, and propane engines.

About 22,800 of the Model 620 units were manufactured before it was replaced two years later by the Model 630. As a group, the John Deere 20 and 30 Series tractors are highly treasured. These were the green-and-yellow tractors that Dad used when the baby-boomers were going off to grade school. They were the last of Deere's two-cylinder tractor production in North America. Many are still working as chore tractors on small farms and are too busy at real work to be available for the collector market.

This 1956 John Deere Model 60 is owned by Marshall Mollett of Greenville, Illinois. This tractor was pulled out of a fence row in the 1980s and was restored.

Here is a front view of a 1959 John Deere Model 630 tractor at East Greenville, Pennsylvania. It was rescued a few years ago from its work as snowplow in Maryland and fully restored.

Model 630

The Model 630 was one of seven 30 Series tractors introduced during the first week of August 1958 by Deere & Company for the 1959–1960 model year. The build date for the first Model 630 unit was August 5, 1958, just a day after the first John Deere Models 730 and 830 were completed at Waterloo. A broad range of improvements had been presented in the 20 Series tractors. For the 30 Series, engineers concentrated on operator comfort and convenience. They also gave it a new, brighter, and more stylish look.

On the surface, designers improved the paint scheme for the 30 Series. They removed the horizontal green rib at the bottom of the hood. Now the bright yellow paint could bleed down and across in a sweeping, simple, eye-appealing form that also required fewer creases and bends on the assembly line. The traditional, round pipe muffler was replaced with a new oval muffler that produced a more quiet and less harsh sound.

The 30 Series tractor introduced new fenders that offered more protection from mud, dust, and accidental contact with the rear wheels. Mounting and dismounting became safer and easier with new fender handholds and a step mounted on the rear axle. Operators sat comfortably, thanks to a deep-cushioned, adjustable seat with a cushioned backrest. The tractor had dual sealed-beam lights mounted in each fender. Inner lights projected light ahead of the tractor and the outer lights flooded the working area.

Steering was redesigned and the steering column now projected upward from the instrument panel to provide a tilted wheel that was more comfortable to grasp. The instrument panel was redesigned with a two-tier approach that presented all the vital operating information on the angled upper tier. Gauges for fuel, oil pressure, ammeter, and coolant temperature were located in an easily read cluster around the steering column. The lower panel held a speed-hour meter, cigarette lighter,

The rear corner view of the Model 630 shows advances in hydraulics and implement control. This tractor is a third generation descendant of the 1934 Model A tractors.

ignition switch, and starter button. Both panels were illuminated for work after sunset.

Starting the tractor was an easy task. The starter pedal on the floor had been replaced. A new solenoid was mounted on the starter and a button on the dash was linked to the solenoid. To start the tractor, operators just pushed the button next to the ignition switch.

The brake pedals were larger and easier to engage while wearing oversized boots. New, flat top fenders were offered, complete with a convenient handgrip and more protection for the operator. On the flat top fenders, a farmer could mount a set of four 12-volt lights. A radio could even be mounted on the fender.

Internally, the 30 Series engine and powertrain were nearly identical to the 20 Series. None of the 30 Series tractors were ever sent to the Nebraska Tractor Test laboratory. They had the same engines as the predecessor and were permitted to use the certification from the test station in line with 20 Series test results.

Both Model 620 and Model 630 tractors operated with six-speed transmissions in a speed range from 1.5 to 11.5 miles per hour. Depending on the fuel, drawbar horsepower ranged from 32.66 horsepower for the All-Fuel engine to 44.16 for the gasoline engine and 45.78 for the LP engine. The end-of-the-line in mid-size row crop tractors now weighed 5,858 pounds, although the LP equipment put another 300 pounds of load on the scale.

Model 630 row crop tractors often were offered alongside a 630 Standard. Excluding tractors built in Mexico, Waterloo built slightly more than 18,000 Model 630 tractors. Nearly 15,300 were in the tricycle or row crop configuration with gasoline engines. Other variations are rare to very rare. The rarest of the group is the 630 Hi-Crop. Only 19 models were built including 11 with gasoline engines, 5 All-Fuel, and 3 LP-gas.

With style and elegance, the Model 630 marked the end of a production era dating back 25 years to the Model A.

This is a 1929 John Deere GP on steel wheels that is owned by Marshall Mollett of Greenville, Illinois. Decades of rust were removed before the restoration began in the 1990s. The lugs are ready to attach for when it goes on exhibition.

Notice the hitch on this 1929 GP. Imagine attaching an implement to it!

The GP standard was followed by the GPT (tricycle) on August 18 and the GPWT (wide-tread) on November 18. The GP family was successor to the Model C row crop tractors. It was still a work in progress.

From 1928 to 1935, Deere & Company built enough Model GP variations to form the basis for a large personal tractor collection. Variations to the GP standard include the GPT, GPWT, GP Potato, and GP Orchard. Each subgroup had several wheel combinations.

The John Deere GP tractor began showing up on dealership lots as an alternative to the Model D. It had standard tread, but it was easy to see this was a tractor with a different purpose. The Model D front axle was attached directly to the engine block. The front wheels were held in a separate frame that provided more accurate steering control. This standard-tread front wheel configuration continued in production through 1935, but other configurations were introduced. The standard GP had a three-speed transmission with a top speed of 4 miles per hour. It was close to the Model D in size, but weighed about 500 pounds less. Both models burned kerosene.

About 70 GPs were built in 1928 and 1929 with the tricycle configuration. The narrowly spaced front wheels rode in the center furrow, while the rear wheels rode in outside furrows. When the engineering was nailed down in the next decade, this became the standard for tractor configuration.

Early GP tractors coming off the 1928 assembly line broke new ground in several ways. In addition to the new steering system, they were the first full production tractor that offered four sources of power: belt, drawbar, power take-off, and an optional new power lift that could be ordered with the tractor.[8]

Power lift was gear-driven. When it was engaged, the engine would provide power to lift or lower a horizontal shaft at the back. The operator could engage the power lift with his toe. The tractor did the work of raising or lowering cultivators, planters, and other equipment.

The GP also had its faults. The L-shaped flathead two-cylinder engine generated less horsepower on the GP than it did on the Model C. At only 20 to 22 horsepower, it struggled with the basic intent for the tractor—a three-row planting and cultivating system. The system was discontinued with the GP family in 1935. Tractor breakdowns were frequent because the air intake plugged easily at first. Deere corrected the problem with a vertical air intake stack in 1929.

Engineers were continuously improving components of the GP. One example is the crossover standard, an interesting variation on the 1930 GP standard. This was a set of GP tractors with a larger six-inch engine bore and improved intake/exhaust system. Most were shipped into Ontario and Manitoba, Canada.

Historian J. R. Hobbs writes of the GP, "...this is one instance where no GP parts book known to be published will tell you anything about these tractors... The

[8] The John Deere Unstyled Letter Series, by JR Hobbs, page 82

This is a 1929 John Deere Model GP tractor owned by Harms Farms of St. Joseph, Illinois. It was a one-owner Kansas tractor until the Harms family purchased it. The tractor was restored in 1996 by sons Derek and Dirk, with their dad's help. It has a front and rear PTO and a power lift.

discovery of some early service bulletins in the late 1980s, as well as the recovery of some of the tractors themselves, proves without a shadow of a doubt that these critters were actually built. Very few of these 68 tractors are known to exist today."[9]

The improved engine became standard for 1931 to 1935. These models are known as the Big Bore GP, as opposed to the Small Bore GP. In the 1931 Nebraska test, the later GP engine burned distillate.

The Big Bore GP engine had cleaner air to breathe and cleaner oil for lubrication, thanks to an improved oil filter. Engineers made changes in the main bearings, governor, steering, final drives, and main case. The front and rear wheels were heavier. Halfway through production for 1931, there were further changes. Engineers increased the cooling capacity with a larger radiator core and brought in a new insulated fuel tank.

Pneumatic or air-filled rubber tires were offered for the first time in late 1932 as an option on the GP. Hard

Two collie dogs wander about the Model GP tractor going back and forth while operating the No. 2 hay stacker. Stacking loose hay for outside storage was an art form in 1929 when this photo was taken.

rubber tires had been an option since the 1920s for industrial use, but this was a new option for the general market. It took extensive testing and marketing, but eventually Firestone Rubber & Tire Co. showed that

[9] The John Deere Unstyled Letter Series, by JR Hobbs, page 85

An extremely rare Model GP 325 three-row corn drill is shown here with the Harms Farms' 1929 GP.

inflated rubber tires offered better traction than steel, not to mention a more comfortable ride.

According to records with the Two-Cylinder Club, 21 separate part numbers exist for GP standard wheels for the front and rear. The basic options include flat or round spokes on steel wheels, round spoke wheels for hard rubber tires, and round spoke wheels for pneumatic tires. Other options were a flat steel wheel with flat spokes (in two sizes), a flat spoke skeleton wheel (in two sizes), an extension wheel for hard rubber, and a wheel with an offset hub for a bean axle.

By 1934, the GP was a much better tractor and the world's economy was looking a little better. Production climbed to 1,250 units. For that year, the GP received a new carburetor. The DLTX-5 replaced the fussy Ensign K carburetor.

As production was winding down for 1935, Deere made one further change. A new Vortox oil bath air cleaner was introduced on the last models. This subgroup of the Big Bore GP series is a highly limited edition with only about 230 tractors in the production run. The final GP standard was finished on February 28, 1935.

The GP standard was the most popular of the 34,500 Model GP tractors. About 30,000 were in the GP standard configuration and 22,000 of those were built in 1929 and 1930. The Great Depression began in late 1929. Production started to decrease and collapsed to less than 170 GP tractors in 1933. Sales began to increase in 1934, but it was too late for the Model GP.

GPWT

Fourteen General Purpose Wide-Tread (GPWT) tractors were built at Waterloo in 1928, beginning on November 12. About 5,100 GPWT tractors were built before the series ended on November 1, 1933. In those five years, the GPWT owners had a selection from 12 wheel options and three different carburetors.

The GPWT had the same engine and transmission as the GP, but it was 2 inches taller, 5 inches longer, and 25 inches wider at the back, hence the name. The later GPWT, introduced with overhead steer on February 10, 1932, was a foot longer. It measured 129 1/8 inches long x 85 3/4 inches wide x 61 1/2 inches high. It is a challenge to load on most trailers because of its size.

This is a Model GPWT with a No. 400 cotton and corn planter. The GP wide-tread version emerged in 1929 to hush the objections of farmers who did not like the three-row concept with the tractor straddling the center row. It had longer rear axles for straddling two rows and a narrow front.

The Wide-Tread was built for three-row crop work, but it could be used for two-row or four-row operations. In good working conditions, it could cultivate up to 45 acres a day when equipped with four-row implements and power lift.

Like the GP, the 1929 GPWT had issues that affected its production life. Steering was very poor and wore quickly, which made it difficult to stay between rows. The air cleaner was prone to plugging until a vertical air stack was adopted toward the end of 1929. A revised cylinder head was provided in 1930 as a temporary fix for problems with the original water-injected flathead engine. Later that year, a new six-inch bore engine was fitted along with a new intake, exhaust manifold, and carburetor. The 1931 model changes included a new location for the air cleaner, the introduction of an oil filter to prolong engine productivity, and an improved system that directed tractor exhaust gases away from the operator. It had improved so greatly that the Union of Soviet Socialist Republics purchased 27 GPWT tractors in 1931.

Further changes in mid-1932 led to a new nickname, the "overhead steer" GPWT. This version had integrated the steering box with the front pedestal, which was now mounted on the front of the tractor chassis. The front wheels also had casters that eliminated wheel

A rare 1930 John Deere Model GP Wide-Tread tractor built for potato farming. This GPWT-P is one of 202 delivered and is owned by Homer Kolb of Phoenixville, Pennsylvania. It has the small-bore engine with the intake stack coming though the hood.

whip when changing rows. Operator visibility and comfort was significantly improved as well. The operator's seat was moved about a foot forward and 10 inches higher for a much better view. Deep in the years of the Depression, production was slow. Waterloo built only about 270 improved GPWT tractors in 1932 and about 150 more in 1933.

This is a detail of the GPWT-P with the brass Ensign carburetor beside the exhaust pipe. The red button on top of the hood designates the filler for the small gas tank. The engine started on gas and then switched to distillate.

Above the belt pulley on this side of the engine, the aluminum-colored R2 magneto made for "John Deere Tractor Company Waterloo, Iowa" by Fairbanks & Morse is visible.

The rear wheel spacing on the potato version was 68 inches, which was 6 inches less overall than the standard wide tread in order to fit the potato rows. Under the axle on the lower left side, you can see the power lift and PTO drive. The spoked wheels and rubber tires are temporary until the original all-steel wheels are restored.

A GP for potato growers was built for one season, starting January 30, 1930, and ending that August. The 202 GP Potato Special tractors had been approved for production by Deere in November 1929. They had a 68-inch rear tread, which was wider than the GP standard by 8 inches and was suited to two 34-inch potato rows. The GP Potato Special also had a new serial plate that started at P-5000. There were two production runs and each lasted about four weeks. Serial number P-5202 was the last unit of the Potato Special. Most were sent to New York, Maine, and eastern Canada. These serial number tags were the only ones Deere ever produced that began and ended with only four digits.

A GP for orchard growers was introduced the following year, on April 2, 1931. It had a four-year run and about 715 GPO tractors were built. Deere had lowered the operator's seat to give the tractor a low, streamlined profile for moving under branches and around tight corners. The GPO benefited from many changes already in place for the GP family. The carburetor was upgraded about midway through production. Deere offered three wheel options for the front and three for the rear. It had an extensive list of other options, such as cast disk front wheels and a concave steel plate that kept low branches from getting caught in wheel spokes and citrus fenders. It proved to be a good, reliable tractor and many have survived. They were shipped to orchards all over the United States and Canada.

Model B

Six months after the introduction of the mid-size John Deere Model A row crop tractor, Deere & Company built the first of a similar but smaller line, the Model B. It was a great success and with 331,000 units built, it slightly edged out the Model A's production line of 320,000 units. There also may be more of the Model B tractors available today than any other John Deere two-cylinder tractor. Model B production peaked at 21,248 tractors produced in 1937. The B weighed about 2,800 pounds, which was 1,000 pounds less than the Model A and 1,600 pounds less than the Model G that was introduced that year. It

also had less horsepower, but it was ideal for tens of thousands of small farms across America. Production ended on June 2, 1952, which was four weeks after the Model A line shut down and about seven weeks before Model 50 was ready to roll.

General categories for the B family included unstyled, early styled (1939-1947), and late styled (1947), which began in the middle of the 1947 model year. Three unstyled variations on the Model B were the BN, BW, and BH. The production of the variations in 1937 included about 400 BN tractors, 88 BW tractors, and only 10 BNH tractors. Most operated with the distillate, the cheapest and lowest octane fuel, and were promoted as an All-Fuel tractor. A gasoline engine also was available and many of these still exist. There were also three purpose-built subsets in the Model B family that are rare to extremely rare.

A basic Model B row crop tractor that is unrestored is about as ideal a starter tractor as any collector can hope

The Model B tractor came out in 1935, a year after the Model A. This is one of the earliest versions with the distinctive four-bolt pedestal. It is pulling a No. 10 corn picker near Dell Rapids, South Dakota.

This 1935 John Deere Model B tractor, parked in a soybean field, is owned by Ken Herbord of Pana, Illinois. It was restored by Herbord in 2004–2005.

Stunning flat-spoked rear wheels make this 1935 Model B outstanding in its field. The original steel wheels were replaced with rubber tires during the restoration.

to find. It should be fairly easy to find and reasonably priced in comparison to any other John Deere two-cylinder tractor in similar condition. Its parts are readily available. The selection of high quality reproduction parts increases every year. The size is right and it is about two-thirds as large as the Model A. It is small enough to fit

nicely in a one-car garage or to haul with a pickup truck. It probably will be a simple tractor to work on. It isn't hard to start and once it's running, it probably won't quit until it's out of fuel.

Unstyled B

The B was, in essence, a scaled down Model A. Commissioned in 1933 for production, it was to be two-thirds the size, power, and weight of the Model A, which was older only by a few months. The first B was built on October 2, 1934.

Other Bs were in the public thoughts that week. On September 30, 1934, Babe Ruth played his final game for the New York Yankees, and President Franklin D. Roosevelt dedicated Boulder Dam. *Modern Mechanix* was introducing a new binding to readers, the coil spring,

This little 1937 Model B on rubber tires is owned by Jim Daly of Sherman, Connecticut, and was restored by Lou and Steve Tencza of New Milford, Connecticut. It has the factory Fairbanks & Morse magneto and a two-piece split rim for the tires.

for a new type of memorandum book that soon would be found in every school bag and desk in North America.

In February 1935, Deere introduced its first significant variation to the Model B to correct a design flaw. The front pedestal had been fastened to the main frame by only four bolts. Field work put too much stress on the four connecting points and led to breakage. Deere solved the problem by giving the B a cast-iron front pedestal securely mounted with eight bolts. Since Deere had built just over 2,000 units, and most four-bolt B pedestals eventually broke requiring a replacement to an eight-bolt version, the four-bolt B is one of the most sought-after collector tractors.

Within the four-bolt group, there are subsets that are the rarest B models. Anyone owning a 1934 or early 1935 Model B should search the serial number records held by the Two-Cylinder Club. The 1934 and 1935 Model B tractors had a brass tag serial number plate intended for the GP series, but they bore the B serial number. Other subsets in the earliest Model B tractors included the first seat frames, which were a solid casting.

Notice the direct steering that is located over the hood. This 1935 Model B has the brass plate for the serial number, which is visible beside the flywheel cover.

A styled 1941 Model B was matched for looks with the Model A. Size-wise, the B was a little smaller than the A.

The second seat frame had a center hole in the casting and the final frame had a reinforcing ring around the shaft to prevent it from breaking. The three types interchange, but they are distinctly different.

Another characteristic of the early unstyled B (and unstyled A) was a fuel cap located under the steering rod in the center of the hood. It forced most owners to use an offset funnel to get fuel into the tank. On the dash, the early unstyled B had an oil gauge as standard equipment. The temperature gauge was an option for a few years.

A version for vegetable gardens, the Model BN, was almost immediately available. Waterloo built about 800 Model BN tractors between 1934 and 1936. Most had the eight-bolt pedestal. The BN was tall, wide at the back, and equipped with a single front nose wheel for vegetable growers to maneuver in very narrow rows. It has been described as a 1930s version of a Stealth bomber. Drivers had to be careful. If they hit a bump with the BN's single front wheel while heading down-hill, they could be launched right off the tractor. In the

This little 1937 Model B on rubber tires with spoked wheels is out in the pasture with its vintage wagon. It's owned by Jim and Pam Brown of Schwenksville, Pennsylvania. Brown did a complete restoration on it in the early 1990s. The wheels are original French & Hecht. The wagon is a McCormick-Deering from the 1940s.

Farm boys discovered that the family's newly styled Model B tractor was easy to operate. This is one of the early styled tractors that were equipped with front and rear steel wheels.

first months, Waterloo built 24 four-bolt BNs that are usually called garden tractors. The four-bolt BNs were shipped to Arizona and California, and they are among the rarest of all John Deere tractors.

A second variation, the BW, was introduced in 1935. It had an adjustable wide front end that was intended to provide more stability in narrow row crops where growers wanted front and rear wheels in the same track. About 250 unstyled BWs were built. A change to a better front axle design made the first 25 of these especially collectable. On the 1936 assembly line, six models were designated as the experimental BW-40. They were given special front and rear axle housings that could be narrowed to 40 inches with steel wheels. Only three are known to survive.

In 1936, two more variations were offered. These were the smaller version of the Model A series AR or standard-tread tractor. They were named the BR and BO. Deere modified the Model B main case to accept a new steering system. Many other changes were made. The end result was a small, fairly popular version of the AR and AO. To adapt the BR to orchard requirements, the BO was provided with differential brakes (so it had a shorter turning radius); a lower air stack; and shields for the air stack, fuel, and gasoline filler caps. As other improvement came along, they were picked up for the BR and BO.

The BR and BO had several important changes in 1939. First, they were given a larger engine to increase the displacement from 149 to 175 cubic inches. With a

A Model BN cultivates beets with a mounted beet and bean cultivator. This tractor was photographed near Phoenix, Arizona, during March 1938.

recalibrated carburetor and other minor small revisions, the new BR and BO had an important power boost. Late in the year, electric starters and electric lighting were offered as options. The models were not restyled with the rest of the Model B family in 1939. Instead, they continued coming off the assembly line through early 1947. Deere built 6,400 BR tractors and nearly 5,100 BO tractors. Out of this group, Deere removed approximately 1,675 for conversion into crawler tractors by the Lindeman Power Equipment Company of Yakima, Washington. Instead of wheel assemblies, the BO main case rode on Lindeman's tracks and undercarriage.

From 1936 to 1941, Waterloo built a Model BI industrial version, based on the standard-tread BR design. The yellow BI was built heavier and had a short BO air stack, heavier and shorter drawbar, modified front end for mounting a blade or snowplow, and was more comfortable. Accessories included a side power take-off, automatic coupler, push plate, a cab, electric start, and lights.

Late in 1936 and early in 1937, some of the Model B tractor family was equipped with a modified radiator. The radiator curtain assembly was replaced by radiator shutters, which made it easier to control engine temperature.

In June 1937, an additional five inches were added to the new General Purpose Model B tractor frame. This increase gave it the same wheelbase as the Model A. Other adjustments were made. The new, long frame B

This unstyled 1939 John Deere Model BO orchard tractor is owned by Homer Kolb of Phoenixville, Pennsylvania. This photo dramatically profiles the standard tread.

series had a longer hood, steering shaft, fan shaft, gas tank intake, and exhaust pipe. The two types are known as short frame and long frame tractors. The easiest way to detect a long frame B is to take a quick look at the

continued on page 69

THREE DEERE PRESIDENTS

Three presidents presided over Deere & Company during the two-cylinder tractor era.[5]

William Butterworth: President 1907-1928, Chairman 1928-1936

William Butterworth married Charles Deere's daughter Katherine in 1892 and joined Deere & Company as an assistant buyer later that same year. The Ohio-born Butterworth was astute, well educated, and a graduate of Lehigh University with a law degree from the National University Law School in Washington.

In 1897, Butterworth was elected treasurer of Deere & Company. As a result of the death of Charles Deere ten years later, he was propelled into the role of president. This new assigned role did not come as a surprise because Butterworth had been informally considered Deere's second-in-command.

Butterworth guided the company through many changes. In 1910, Deere's board of directors launched a major reorganization to unify the 11 factories and 25 sales organizations throughout the United States and Canada into one consolidated entity. By 1912, the modern Deere & Company had emerged.

The company's product line continued to grow under Butterworth's leadership. Deere entered the combine harvester market in 1912 and purchased the Waterloo Gasoline Engine Company in 1918. With the latter acquisition, Deere gained the rights to manufacture and sell the popular two-cylinder Waterloo Boy tractor, which was the company's first product in the tractor market.

Butterworth was elected Deere's first chairman of the board in 1927. He was elected president of the United States Chamber of Commerce four times before his death in 1936.

Charles Deere Wiman: President 1928-1955

Charles Deere Wiman was the son of William and Anna Deere Wiman, the great-grandson of John Deere, and the nephew of his predecessor, William Butterworth.

Wiman began his career at John Deere in 1915 as a line employee with a salary of 15 cents an hour. As he advanced within the company, he gained the respect of shop employees, engineers, and supervisors.

His career combined dedication to the company with loyalty to his country. Wiman took several leaves of absence from John Deere to serve in the military. In 1916, he trained as a civilian pilot, received his license, and flew for three months before he was injured in a fall. After his recovery, he returned to Deere & Company until the United States entered World War I. Wiman was commissioned as a U.S. Army second lieutenant. He reached the rank of captain before being decommissioned in 1919.

He was elected to Deere & Company's board of directors in 1919 and he became vice president in charge of factory operations in 1924. With the retirement of William Butterworth in 1928, Wiman became president of Deere & Company at the age of 35.

Wiman placed an emphasis on expanded research and new product development, even during economic downturns. In 1934, during the peak of the Depression, John Deere introduced the famous Model A tractor. The Model B followed the next year. Both tractors were highly successful and remained in production until 1952.

World War II brought Wiman's re-entry into the military. In 1942, he was commissioned as a colonel in the U.S. Army and attached to the staff of General Levin H. Campbell, Chief of Ordnance. In 1944, Wiman was appointed director of the

---

[5] http://www.deere.com/en_US/compinfo/student/past_leaders.html

Farm Machinery and Equipment Division of the War Production Board. To carry out his duties, the Army placed him on inactive status. He rejoined Deere & Company in mid-1944 and remained at its helm until his death in 1955.

William Hewitt: President 1955-1964, Chairman 1964-1982

William Hewitt was the last Deere family member to lead the company. Born in California, Hewitt received a degree in economics from the University of California at Berkeley. He attended Harvard Business School for a year, but lacked the funds to complete a graduate degree.

After serving in the Navy for four years, Hewitt joined the Pacific Tractor & Implement Company. In 1948, shortly after his marriage to Charles Deere Wiman's daughter Patricia, he reluctantly joined Deere & Company as a territory manager in California.

Hewitt quickly gained a reputation as a good leader and a quick learner. He was known for his ability to remain cool under pressure. He was named a director in 1950; in 1955, he was unanimously elected to the newly created post of executive vice president. Twelve days after his father-in-law's death in May 1955, Hewitt was elected Deere & Company president. Hewitt was a very active, hands-on executive.

During Hewitt's tenure, John Deere took its first steps toward becoming a multinational corporation. In 1956, the company purchased a majority share of the Lanz tractor factory in Mannheim, Germany, and obtained land in Monterrey, Mexico, for a new tractor facility.

Hewitt managed the process of bringing the new four- and six-cylinder tractors into the Deere line. Seven years of development culminated in a spectacular introduction of the models, which were called the "New Generation of Power," at a noteworthy 1960 dealer show in Dallas.

The company continued its expansion around the world for the next two decades. By the 1970s, Deere & Company had a presence in Argentina, Venezuela, France, Spain, Great Britain, Western Europe, Australia, Japan, South Africa, China, and the Middle East.

Hewitt spearheaded the construction of a new corporate headquarters for the company. After he decided that the company's center of operations should remain in Moline, Illinois, Hewitt selected renowned Finnish architect Eero Saarinen to design the building. The striking new Deere Administrative Center opened in 1964 and has been cited as one of the world's best architectural designs.

William Hewitt retired in 1982 and he was named U.S. ambassador to Jamaica that year. He died in 1998 at the age of 83.

Continued from page 67

air horn on the carburetor. On a short frame B, the air pipe makes a straight vertical joint between the carburetor and air cleaner. On the long frame unstyled B, it is horizontal for about five inches and then has an elbow to take it up to the air cleaner.

Late in 1937, Deere met requests for variations in the BN and BWH. Growers wanted a high clearance BN, so Deere added two inches of vertical clearance and called the new version BNH. Owners also could adjust the rear tread to 104 inches wide. Only ten were built in 1937 and another 55 were built in 1938, prior to introduction of styling for the Model B family. In 1938,

Deere built 51 unstyled BWH tractors. These were the wide front version of the high clearance line to provide more stability.

One more variant, the BWH-40, was produced in 1938 but the production numbers are unknown. It had a minimum tread width of 42 5/8 inches and required fenders as standard equipment. It was available only with rubber tires. It was intended for crops grown in 20-inch rows on 40-inch beds, but the tread could be spaced up to 80 inches wide. At least six tractors exist.

Generally, a Model B could be ordered with wheel equipment to suit the owner. Rubber tires were available from the start. Wheels for rubber can be found with

A styled Model B cuts the corn crop. This photograph was taken in 1948.

round spokes, flat spokes, and cast-iron centers. Flat steel rear wheels and steel front wheels with guide bands were standard. Spade lugs could be obtained in two sizes. There were specialty lugs (sand, button, cone), and for growers with sticky soil, the skeleton-style steel rear wheel was an option. A very rare option was a 1937 tip-toe-style steel rear wheel.

Styled B

On June 16, 1938, the John Deere tractor assembly lines completed production of the first Henry Dreyfuss-styled Model B tractor. It was a 1939 model and the series was an immediate success. The Model B became the tractor sales leader for Deere & Company and held that position through the end of the 1946 model year production.

There was unmatched graceful elegance to the exterior 1939 Model B. Dreyfuss had started the process of designing features for operator visibility, comfort, convenience, and safety. This two-cylinder tractor looked better and was a nicer place to work. His ideas would filter through the tractor parade at Deere dealerships.

Inside, the two-cylinder engine had more cylinder displacement to give it more power and more torque. The styled B had a wide variety of wheel equipment options and a choice of factory-installed transmissions. Fenders and hydraulic power lift were optional. The

This John Deere BO crawler shows the distinctive Lindeman name on the undercarriage. It was built in late 1946 and is owned by Paul and Robert Watral of Hauppauge, New York.

styled B buyer could also order an electric lighting package and electric starting.

The styled B was available in a wide variety of configurations. The most popular was the B tricycle model. There was also the BN with a single front wheel; the BW with wide adjustable front axle; the BNH that combined a single front wheel with higher clearance; the BWH that combined high clearance with wide,

This styled 1948 John Deere Model BN, with its single front wheel, fits perfectly in the corn rows. It's owned by Frank Swotkewicz of Mattituck, New York. It had a major rebuild after it was purchased in 2003. Features on this tractor include electric start, six-speed transmission, and an original complete Yakima 801 three-point hitch.

adjustable front end wheel spacing; and the BWH-40, which was a version with special narrow tread for the rear axle and had high clearance and fenders.

More big changes came along for the 1941 model year. There was a further increase in horsepower, an improved carburetor, and a new six-speed transmission. Rubber tires had become very popular. The tractor and operator could handle a little more road speed with rubber tires. Deere enabled that speed increase with the new transmission.

Few changes occurred during the war years. The styled B dash on tractors with electric starting was upgraded for 1941. As the war was coming to an end, Deere & Company introduced the first system that could operate a remote hydraulic cylinder on a farm tractor in October 1945. Powr-Trol was exclusive to Model B tractors for a short while, due to limited production. When Powr-Trol went into full production in the 1947 model year, it became a retrofit for many of the earlier tractors.

Here is a spectacular fall day and a spectacular 1939 John Deere Model BR, restored and owned by Pete Loman of Orange, Connecticut. This tractor still has its original rear tires.

One of the later styled Model B tractors is raking a hay crop into swaths. This tractor was photographed in 1951.

Late B

Production for the late styled Model B began on February 4, 1947, and ended on June 2, 1952. Production included approximately 96,000 tractors with gas engines and 13,000 with All-Fuel engines. Deere & Company reduced the versions to just three: the B, BN, and BW. The BN and BW were first built on March 21, 1947.

The early B was good, but the late B was greatly improved. It was still rated as a small row crop tractor, but it was 500 pounds heavier and had much more under the hood. The revised engine became known as the Cyclone. Engineers had placed an "eyebrow" next to the intake valve to induce an intake swirl effect. This kept the air-fuel mixture in suspension to enable more power to be produced. Engine bore was increased as well to give the new engines 190 cubic inches, and engine speed was cranked up to 1,250 rpm from 1,150 rpm. Rated horsepower jumped from 20 to 27, and maximum pull increased from 2,756 to 3,437 pounds.

A new, six-speed transmission with a creeper gear and a single control lever was also introduced. Electric lighting and the electric starter became standard features. A cushioned seat was introduced and operator controls were more convenient than ever.

Up front, the 1947 late styled Model B rode on Deere & Company's new patented knee-action suspension system. The Roll-O-Matic suspension offered a much smoother ride. It is similar to a semi-independent front suspension. When one wheel hit a bump, a gear

A styled Model B is shown operating an early hay baler. This photo was taken around 1950.

would force the other wheel down to keep at least one wheel in contact with the ground. The improved suspension prevented the tractor's nose from hopping, which made the ride more comfortable, as well as easier to control and steer. At the rear, Powr-Trol became standard and Power Lift was available. Other popular options included fenders and heavy cast rear wheels.

Once the late B series was introduced, only a few changes were required. In 1948, the Wico C magneto was replaced by the Wico X. In 1949, a one-piece brake shaft made it easier to service the brakes. In 1950, the Wico XD distributor was adopted and became standard. Buyers were also offered a two-piece front pedestal in 1950. A

This 1953 John Deere Model 50 is owned by Phares and Lois Stauffer of Livonia, New York. The original Roll-O-Matic front end was replaced, for safety reasons, with the wide front end from another John Deere.

new Roll-O-Matic was offered for the two-piece pedestal and provided a choice of four front-ends. The rear axle housing was adjusted from round to square, and clamshell fenders replaced the earlier design. During the Korean War, copper was in short supply, so the last Model B series tractors had steel radiator cores and came with a water pump to make up for the less efficient heat transfer.

Model 50

The first John Deere Model 50 row crop tractor was built on July 22, 1952, and stayed in production until May 14, 1956. Historically, on July 22, 1952, California residents were surveying damage from a 7.8 earthquake that had happened the day before in Kern County, and the 15th modern Olympic Games were underway in Helsinki, Finland.

The Model 50 was Deere & Company's third introduction of Number Series tractor replacements for row

crop tractor families from the Waterloo factory. Three months earlier, the first Model 60 and Model 70 tractors had replaced Models A and G.

In a policy change, Deere & Company discontinued building and naming front-end variations at the factory. Instead, the manufacturer built and shipped five front-end choices to the dealerships. The front ends were installed by dealers according to buyer preference. The front-end choices were single wheel, dual wheels, Roll-O-Matic wheels, fixed 38-inch tread, and adjustable-tread axles.

Standard Model 50 rear tread was adjustable from 56 to 88 inches. The standard wheelbase was only 90 inches, the turning radius was 103.5 inches, and shipping weight was 4,435 pounds. The Hi-Crop option was not offered for the Model 50, although it was available for Models 60 and 70. An optional, extra-wide rear axle also was available for three row crop tractors from Waterloo.

Notice the No. 800 three-point hitch in this rear view of the Model 50. The tractor is equipped with an aftermarket power steering kit.

The Model 50 had an awkward first season on the production line. After the first five were built in July, the annual factory shutdown began. Production resumed in August, but stopped again after another 125 were built due to a steel strike that started in October. Gasoline and All-Fuel engines were available from the start for Model 50.

The gasoline engine had a new, dual carburetor and was tested at 47.5 horsepower on the drawbar.

The two-piece convertible pedestal became standard on the Model 50. It allowed front-end configurations to be made or changed at either the dealership or the farm level. It was specific to the selected type of

Only for show, this 1957 John Deere 520N row crop tractor is owned by Ralph Knabb of Spring City, Pennsylvania. This tractor was shipped on November 30, 1956, to Syracuse, New York, and is equipped today with front-mounted rockshafts, front and rear wheel weights, and an external air pre-cleaner. Everything but the transmission was rebuilt, repaired, and brought to factory specs in 2006. It has operated less than 3,600 hours in its lifetime.

configuration. Other changes in 1954 included a new Delco-Remy distributor and a conventional thermostat system to replace the new and troublesome thermostatically controlled shutter system.

Power steering improved in 1955 when Deere & Company introduced the first fully integrated, factory-designed, power steering system for farm tractors on Model 50. It was an engineering breakthrough and a major advance in operator comfort and convenience. It was an optional system and it was upgraded later, but power steering took the heavy muscle work out of steering a tractor while loaded with front-mounted equipment in soft or muddy field conditions. Front-end accessories included loaders, cultivators, and corn pickers. Others included an hour meter, wheel weights, and frame weights.

In 1955, on request, Waterloo shipped a few Model 50 tractors equipped to burn liquid propane. Eventually, dealers sold more than 700 of these LP-equipped Model 50 tractors. In 1956, its final produc-

tion year, Model 50 was equipped with the 801 Traction-Trol form of the three-point hitch. It replaced the 800 and 800A hitches.

As a footnote, it should be mentioned that Waterloo sent 200 Model 50 specials to the Barber-Greene Company of Aurora, Illinois. These models became power units for the Barber-Greene Model 550 windrow loader. They had a special cast 28-inch rear wheel and a single front wheel with a shorter yoke. The first gear ground speed was slashed 75 percent to aid the loader as it built windrows with snow, dirt, and other material. The special didn't have a PTO or rockshaft, and it was all-green. When production ended, Waterloo had built about 32,500 of the Model 50 tractors. They were popular and tough little tractors, and they still are highly regarded.

Model 520

Deere & Company replaced the first Numbered Series from Waterloo with what became known as the 20

This 1956 John Deere Model 520 has the traditional narrow front Roll-O-Matic wheels. This tractor is owned by Schewe Farms of Greenville, Illinois. It was restored by Maurice Schewe in the early 1990s and now is owned by his sons, Mark and Bill Schewe.

Series tractors. They were on the line for two production years beginning in May 1956 and they were replaced by the 30 Series in 1958.

The first Model 520 was built on May 4, 1956, and the last was built on August 1, 1958. The 520 introduction was followed in June by Model 620, then Models 729 and 829 followed in July.

The Model 520 replaced the 50. Like its predecessor, the 520 was offered in a basic tricycle configuration. It was the smallest tractor on the Waterloo assembly line. In the gasoline and All-Fuel versions, it shipped at 4,960 pounds, compared to 5,858 pounds for the Model 620 and 7,800 pounds for the Model 720 that came off nearby assembly lines at the same time.

The Model 50 and the Model 520 shared the same 190-cubic-inch engine. For size, they were in the Model B tradition as Waterloo's smallest row crop tractor. The 520 was a greatly improved and a quite different tractor.

Engineers had found another way to boost power on the two-cylinder engine. The drawbar power with gasoline soared from only 16 horsepower in the original Model B, to 27 horsepower with the Model 50, and then to 33 horsepower with the Model 520. It had a better combustion chamber design, higher compression ratio, and a boost in rpm. The new engine speed was 1,325

rpm. The final drives were beefed up to handle the increased power, as was the transmission.

Buyers had options for an All-Fuel or LP-gas engine, but they voted heavily for gasoline. The LP-gas engine was nearly 300 pounds heavier. The All-Fuel engine with lower octane fuel produced slightly less power. Records show that in two years, only 83 Model 520 tractors were built with the All-Fuel engine and about 750 with the LP-gas engine. The other 13,000 Model 520 tractors had gasoline engines.

The standard Model 520 front end was a dual-wheel tricycle configuration. Dealers could supply options including Roll-O-Matic, a single nose wheel, an adjustable wide axle (48 to 80 inches wide), and a 38-inch fixed-tread axle.

A few items were modified for the 1958 model. An axle-mounted step became standard equipment. The tractor dash was now all black. The tractor had sealed-beam lights and a steering wheel with a plastic cover.

Sales of Model 520, compared to other row crop tractors in the Waterloo 20 Series, were lower. During the same period, Waterloo built approximately 23,000 of the mid-size Model 620 and more than 31,000 of the large Model 720. Sales were probably being drawn away by a slightly smaller row crop tractor at Deere dealerships from the new factory at Dubuque. Sales of Deere's smallest row crop, the Dubuque-built Model 420, reached more than 47,000 tractors in the two-year period.

Model 530

The Model 530 was nearly identical to the late Black-Dash Model 520. The first Model 530 was built on August 4, 1958, the same day as Models 830 and 730, and a day before Model 630. The 530 was mostly a cosmetic upgrade to Model 520. The engines and drivetrains were identical and most components were interchangeable. The big difference is in sheet metal and the painting scheme.

The Model 530 was Waterloo's small, general purpose row crop tractor successor to the Model B. It weighed nearly twice as much and could pull 4,600 pounds versus 1,700 pounds. In addition, it was more

This 1959 John Deere Model 530 is also owned by Schewe Farms of Greenville, Illinois. Notice the updated painting style, which is typical of 30 Series tractors. Under the hood, this tractor is identical to the Model 520.

comfortable, operated with a six-speed transmission, and had gained rubber wheels.

The dual-wheel, tricycle-style Model 530 front end was standard. Other front options included the dual-wheel front with Roll-O-Matic (in regular or heavy-duty versions), a regular or heavy-duty single front wheel, an adjustable wide front axle, and a 38-inch fixed-tread front axle. The entire 30 Series offered creature comforts with appeal today, such as power steering, electric starter, electric lights, and excellent hydraulics.

As with the Model 520, production trailed for the Model 530. It was lowest in the 30 Series row crop

tractors at approximately 10,300 tractors in a two-year production run. Within the 530 family, gasoline engines continued to be most in demand. Approximately 9,800 were supplied with gas engines compared to 417 LP-gas engines and 83 All-Fuel engines.

The Model 530 was the last two-cylinder row crop tractor built by Deere & Company at Waterloo. Production ended on September 27, 1960, the day after Ted Williams hit his 521st home run, the day after Nixon and Kennedy's first television debate in Chicago, and the day after Fidel Castro's record-length speech of 4 hours and 29 minutes to the United Nations.

This is a 1951 John Deere Model GW owned by Doug Roan of Clear Brook, Virginia. It plowed snow in a Minnesota parking lot for four decades before Roan and friends decided to restore it during the winter of 2006 and 2007.

WATERLOO FACTORY
BIG BROTHER AND
LITTLE SISTER TRACTORS
MODELS G, 70, 720, 730, H

By 1934, through invention, trial, and error, the Deere & Company directors, engineers, factories, and workers had worked out the complexities of building a high quality, reliable, affordable row crop tractor. That year, as economic pressure was starting to ease in big cities, the company introduced two row crop tractor families. Sales soared and successors to those first row crop tractors with two cylinders stayed in production at Waterloo until 1960.

Almost immediately, it was evident that the technology could be scaled up and down. Modest improvements could improve the capacity and performance of both Model A and Model B. Directors approved steps in both directions for the Waterloo factory in mid-decade and even opened the door to a utility tractor at the Moline factory.

At the top end, the largest farms could afford and use an even larger row crop tractor. This was remedied in the 1937 model lineup with the introduction of the Model G. The Model G and its successors, the 70, 720, and 730, were built on the Waterloo assembly lines between 1937 and 1960. Total production reached about 163,000 tractors. The market share varied within the production years, but Waterloo typically built two small

Model A tractors to each Model G, and the mid-size Model B tractors were even more popular.

At the lower end, thousands of farms still relied on a single team of horses or mules for plowing, planting, and weeding. Below that, even smaller farms relied on a single horse or mule to put in a few acres every year to feed the family and help with farm chores. The benefit of tractor technology over the horse or mule wasn't in dispute. Tractors in the right price and power range could generate thousands of new tractor sales. Deere & Company came up with a two-stage response.

In 1936, directors commissioned the design of a much smaller utility tractor to replace a single horse or mule. They assigned it to available factory space in Moline and construction of this utility tractor began the following year with Model Y (see Chapter 7).

Shortly after, Waterloo engineers split the difference between what they saw the Moline tractor could be and what they already were developing as an improved Model B. This proved to be a perfect fit for the market and the dealerships. It filled a range of farm needs and provided dealerships with an impressive long line of tractors in many sizes to suit every farm budget.

This Model G pulls a potato planter in New York State in 1938.

Even the big brother Model G was introduced, and Waterloo engineers created the Model H, a little sister to the other row crop tractors. The Model H was introduced in 1938, a year after Model G. It became the fourth and final row crop tractor introduction from Waterloo in the two-cylinder tractor families. The Model H was very successful but short-lived. Production of the replacement was shifted to the Deere & Company new dedicated tractor facility in Dubuque, Iowa, and located beside the Mississippi River. The Dubuque Model M became the replacement at dealerships to the Waterloo Model H.

In June 1938, the Nebraska Tractor Test station saw the arrival of John Deere models A, B, and G. Model G was rated for three 14-inch plows and weighed 4,400 pounds. The Model A handled two 16-inch plows and weighed 3,783 pounds. The lightweight Model B weighed 2,878 pounds and was rated for two 14-inch

plows. Over the next two decades, a lot of steel was added to all three tractor families and the little two-cylinder engine gained speed, displacement, and power. For the final Nebraska test in 1958, the large Model 720/730 Series tractor weight ranged from 6,790 pounds to 8,470 pounds for the Hi-Crop diesel version.

Unlike the smaller row crop tractors from Waterloo, the Model G family took two direct hits during World War II. Deere & Company had planned to introduce a styled G in early 1942. It would be the final row crop family for Dreyfuss styling. A price increase was planned for introduction with the styled G. Those plans went on hold after the bombing of Pearl Harbor on December 7, 1941. One of the first actions of the new War Production Board, which came out about three months ahead of plans for the Styled G, was a regulation that prohibited the planned price increase. Choosing between no price

This Model G is shown pulling a John Deere No. 93 disk plow in 1937.

increase or continuing production of an unstyled tractor, Deere & Company introduced a name change. They cancelled production of the Model G and began building the modernized 1942 Model GM tractor.

The second hit came six months later, in September 1942. Deere suspended production of Model GM tractors. The suspension lasted 24 months. Plans were drawn up for a 1944 GM, but it never got off the drawing board. When production of this family was able to resume in October 1944, the new tractors coming off the line were identified as the 1945 model. However, Model GM production remained slow. Sales were better for the smaller row crop Model A and Model B families.

When the war was over, Model GM disappeared from production lines and was replaced by the Model G. Around the same time, some configuration options were offered for the first time with the styled Model G family. Like the other tractor families from Waterloo, the

Model G was replaced in the 1950s by Model 70, then by the 20 Series, and finally by the 30 Series. The largest two-cylinder row crop tractors were reliable as well as large, heavy, and powerful. They are certainly available for collectors, but they are less common.

Model G

The first John Deere Model G row crop tractor was marked built at Waterloo on May 17, 1937. It was a century since John Deere had built his first moldboard plow. A civil war was raging in Spain and a Fascist political axis was forming in Europe. In America, the economy was recovering. San Francisco's Golden Gate Bridge would open in a few days, and War Admiral was just one win away from horse racing's prized Triple Crown victory.

The Model G was the fruit of a decision to build three distinct sizes of row crop tractors at the Waterloo factory for different markets. The first size, the mid-size Model A,

A matched pair of Model G tractors are equipped with upscale chrome intake and exhaust stacks for parades and shows. Both tractors are owned by Francis Bujnicki.

A 1949 John Deere Model G, owned by Francis Bujnicki of Wading River, New York, is shown here with a serious display of the American flag. This tractor farmed Long Island potatoes for many years and has been restored to factory original condition. It's still used for plowing every spring and appears in local parades.

launched in 1934 with 24 horsepower. The second size, the Model B, had been launched a few months later and was the smallest of the three. With glowing reports and high sales for the first two row crop models, Deere directors knew they had a winning approach. However, it was better to wait for some improvement in the economy before launching the biggest row crop tractor.

The Model G was built for heavy work. It had a totally new, high-torque, two-cylinder engine with a huge 413-cubic-inch displacement. The four-speed transmission and final drives were re-engineered to handle all the new power. On the growing number of large farms, it could handle three 14-inch plows, a 10-foot tandem disk, 3-row bedders under touch conditions, 4-row bedders in most conditions, plus 4-row planters and cultivators. The belt could power a 28-inch threshing machine at full capacity. This kind of power made it popular with large

acreage cotton and corn growers, and with grain farmers who grew a few acres of the common row crops. If well-maintained, the Model G was likely to run forever.

Managers soon learned they had missed one necessary upgrade for the upscaled row crop tractor. Soon after it entered field service, owners were struggling with the cooling system. Pulling heavy loads in hot weather was easy enough for the big engine, but the cooling system would overheat. Engineers quickly solved the problem with a new radiator design. Waterloo started building a new, high radiator Model G on January 19, 1938. The revised Model G had a larger, taller radiator, along with a revised fan and shroud, revised radiator shutter, a larger fuel tank, and a new hood. The top of the tall radiator gained a distinct new groove to provide clearance for the steering shaft. On early low radiator models, the steering shaft clears the top of the radiator by about a half inch.

Approximately 1,600 Model G, low radiator tractors were built in the first season. A field modification program was available through dealers and it reached most but not all of the low radiator G tractors. Overheating was less frequent in the northern states and Canada, so some units remained original. A few privileged collectors have the original, unmodified, low radiator G to prove the story.

The 1939 model year included modifications to the cylinder block and head, which allowed more efficient heat transfer and coolant flow. A new, upper water pipe also was required. By the time the assembly line stopped building Model G on December 22, 1941, many other modifications had been made to the big tractor. In total, less than 10,700 Model G tractors were built before the war stopped production. In addition to many part changes, the Model G also had a significant number of wheel options, including both steel and rubber. They are a simple tractor model that are easy to understand and operate, but they need a big garage or big shed.

Model GM

The styled and modernized Model GM was introduced on February 18, 1942. Sales remained slow in comparison to sales for the smaller row crop tractors. Waterloo built approximately 750 Model GM tractors in 1942 before suspending production due to material shortages. It took about twice the steel and other metals to build a GM as opposed to the little styled B. The 1945 model was on the assembly line when production resumed in October 1944. War in the Pacific had ended weeks earlier and it wasn't long before Deere & Company cancelled the GM series. Waterloo had built approximately 8,800 GM tractors.

Model GM was an excellent, good-looking, two-cylinder tractor for a large farm. It had an improved powertrain and styled sheet metal. The engine ran cooler and had a bit more power than the predecessor in the Model G. It had an improved Marvel Schebler DLTX-51 ("big nut") carburetor. A new six-speed transmission handled by a two-lever shift system was available, although it was in short supply. Shutter controls were

improved. The air intake and exhaust remained side-by-side stack rather than inline, which had been switched on the styled A and styled B.

Electric starting and lighting were optional on the big tractor. Other options included Power Lift and fenders. In October 1945, Deere introduced Powr-Trol as an option for all its row crop tractors, including the GM. For the first time, the tractor could operate a remote hydraulic cylinder for precision positioning of the rockshaft.

The Model GM could make a good project for collectors who specialize in steel wheels. Model GM tractors were built with four variations in steel wheels, but only two are likely to be found in North America. When these big tractors left Waterloo on steel wheels, they also left with only four forward gears. A six-gear transmission was available, but field speed at more than 6 miles per hour on steel was dangerous.

Styled G

The styled Model G family, built for model years 1947 through 1953, is more suitable for collectors than the Model G or GM because it has some diversity. There's an early styled G, a late styled G, and subsets. The supply is best for the late styled G because Waterloo built nearly 32,000 of that variation. On the flip side, the early styled G was phased out after only five months of production and about 2,600 tractors. The early styled G had two versions; the late styled G had three versions. Any tractor in one of the styled G subsets is highly collectable.

The first styled G rolled off the assembly line on March 7, 1947. It looked about the same as the last few GMs that were still being assembled, but it did have only a letter G on the serial number plate rather than GM. The styled G also had the Roll-O-Matic knee action option, which was already a popular choice on the smaller row crop tractors. It had electric lighting and a starter. For the farmers who preferred the old and reliable hand-crank starting and steel wheels in the field, they still could order those options for their styled G.

In June 1947, a Model G and Model A both went through the Nebraska Tractor Test lab. The Model G burned All-Fuel and the Model A burned gasoline. Both engines operated at 975 rpm and produced virtually the same horsepower: 38 horsepower on the brake and 34 on the belt. However, the Model G was nearly 900 pounds heavier and was able to pull an additional 250 pounds of load. It also had more bore and stroke by about 90 cubic inches. The Model G's cylinders were 6.12x7 inches, compared to 5.5x6.75 inches on the Model A.

After the World War II veterans were back home, Deere & Company decided it was time to offer some configuration options on the biggest row crop tractor. The split pedestal already used on Model A and B tractors was retooled for the larger styled G family. The factory began taking orders for narrow, single front-wheel GN versions and wide-axle GW versions. The GW could be ordered with adjustable rear axles, which made the configuration versatile enough to meet a grower's demands. With the 104-inch rear axle, the biggest row crop tractor was well suited for narrow row crops and for crops grown on raised beds.

The late styled G, introduced at the end of July 1947 as a 1947 model, was a bit more comfortable. It had a box-style seat instead of the old pan seat. Batteries were under the seat in a more convenient location. It had a few cosmetic changes and a belt pulley guard to prevent dirt and oil from flying into the operator's face.

The late styled G, GN, and GW tractors had quite a number of minor changes. The parts are not always interchangeable: The camshaft and bearings changed, the PTO shaft was standardized, there was a new distributor, a new water pump, and a steel radiator core. There were numerous wheel options. The G, GN, and GW had six options for rear wheels on steel or rubber, and several versions for front wheels.

For 1951, Deere responded to sugar cane growers in Louisiana and Florida with a Hi-Crop Model G. For them, the 412-cubic-inch GH All-Fuel engine was less expensive to operate than the late styled A and it provided a heavier machine for the work. Over two model years, Waterloo built about 240 of these GH tractors and exported half of them.

What Deere didn't do with the styled G is interesting. It never offered a standard-tread GR tractor, similar to the AR or BR. It also never supplied an LP-gas, diesel, or gasoline version of the G, although some options were provided by aftermarket suppliers. The All-Fuel Model G was not an option because it was the only engine offered. Production ended on February 18, 1953, with a Model GW tractor.

Model 70

March 27, 1953, marked the first build date for Model 70 tractors, which were the successors to the late styled Model G. Production ended on June 22, 1956. The tractor was available in three configurations and with four types of fuel. Going back 16 years, the total production of large two-cylinder row crop tractors at Waterloo had reached approximately 60,000 units. Most years, sales had been slow. The forerunner, late styled G tractors, sold about 32,000 tractors in six years. In just four years, the Model 70 sold more than 41,000 tractors.

The Model 70 was a much better tractor. It was more comfortable, easier to operate, had an all-new engine, power steering, better hydraulics, live PTO, rack and pinion rear tread adjustment, and a 12-volt electrical system. It continued using the same high-low transmission as the Model G and weighed about 6,000 pounds.

By the time it was ready to replace the late styled Model G, Deere & Company had switched to a Number Series designation for new models of two-cylinder tractors. First had been the mid-size Model 60 that replaced the Model A in June 1952. The smaller Model 50 had replaced Model B. The next Number Series up in scale, Model 70, was unveiled in spring 1953 as the latest biggest-and-best row crop tractor, with a choice of two engines: gasoline or All-Fuel.

Like the 50 and 60, the 70 was a very different tractor from its predecessor. It had more power, more weight, and more comfort than ever before. Unstyled early tractors and late 30 Series tractors tend to gain more attention from collectors, but in its day, the 70 was a very popular tractor and a new day may come.

The basic row crop Model 70 had the same All-Fuel engine as the Model G family. It had developed into a very good engine. Thanks to a new dual barrel carburetor

This is a 1955 John Deere Model 70 owned by Phares Stauffer of Livonia, New York. It was fully restored in 2005 with the help of Ralph Knabb.

and improved distributor, the Model 70 produced six more horsepower with almost no backfire issues.

Deere developed a gasoline engine for the Model 70. At 379 cubic inches, it was smaller than the All-Fuel engine, but it also had more power. It had the same duplex carburetor, a somewhat higher compression, and an improved cylinder head design.

A year later, Deere offered a third engine option: LP-gas. The LP system was factory engineered and not an add-on. It achieved nearly 52 horsepower on the Nebraska test, which was the highest of the Model 70 engine series.

A split front pedestal was now standard on the Model 70 and enabled Deere to offer five front-end choices for configuration: standard dual front wheels, Roll-O-Matic front wheels, a single front wheel, a 38-inch fixed-tread front axle, and a width-adjustable

front axle. At the rear, there were options for long or extra-long axles with tread from 56 to 112 inches.

The Model 70 came equipped with Powr-Trol for live hydraulic power and had an option for a live PTO. It had a new No. 800 Series three-point hitch. Factory-engineered power steering became available in 1954 and it came with a smaller three-spoke steering wheel.

The 800 Series hitch became available in 1953 for the new Model 50, 60, and 70 tractors from Waterloo. It replaced the Yakima 2100 Series hitch that had been a popular accessory on the big row crop tractors in the western United States. The 800 was built by the Plow Works in Moline so it had to be ordered and installed by dealers. Deere & Company described it as, "With the separate hitch bar which attaches to the lower draft links, you attach and detach many tools right from the tractor seat…There's no wrestling with the implement, no

Under the hood of the Model 70, Deere & Company promoted its new power steering system. A factory-mounted option for a tachometer is just above the flywheel cover.

need to jockey the tractor into exact position in order to line up the hitch points."[10]

Waterloo configured three units in 1954 as Model 70S tractors with standard-tread front and rear: two with a gasoline engine and one with LP-gas. It was an immediate success. Sales took off and more than 3,000 were built in the remaining two model years. A fourth fuel option, diesel, became available for 1955 and 1956 models with a choice of row crop, standard-tread, or hi-crop configuration. The diesel version was very popular and led to sales of 14,397 tractors in just two years, even though it added about 25 percent to the purchase price.

The Model 70 was Deere's first row crop tractor that burned diesel. It was more economical to operate than the diesel-burning, standard-tread Model R that had just ended its seven-year production run. The Model 70 diesel broke the all-time fuel economy record at Nebraska that had been held by the John Deere Model R. On diesel, the Model 70 cranked out 51.5 horsepower on the belt and 54.7 horsepower on the drawbar. A new center main bearing for the crankshaft helped keep the crankshaft in place while it was under load and provided better balance for the engine.

Owners started this 376-cubic-inch, high-compression diesel with the aid of the new V-4 gasoline-fired starting engine. The little four-cylinder gas engine provided unlimited cranking time for the diesel. Its exhaust warmed the diesel's intake air. The two engines shared one cooling system. The pony engine was started by pushing a lever forward. When the V-4 was running steady, the operator pulled the lever back to engage the flywheel ring gear and turn the big diesel engine. When

10 The First Numbered Series of John Deere Tractors: Models 40, 50, 60, 70, and 80. Hain Publishing, 2004, page 70

Notice the clamshell fenders on this Model 70 and the No. 801 hitch.

Model 70 had an updated, comfortable seat; a new gearbox; and an instrument cluster on the dash, including a cigarette lighter.

the diesel was turning over, the lever was moved partially forward to put the diesel on full compression and operate alone.

The 1956 Model 70 saw several minor changes. The most popular was probably the new 801 Traction-Trol hitch. It provided a mechanical weight transfer to the rear wheels from implements mounted on the hitch to give more traction. Many of the earlier 800 and 800A hitches were replaced with the 801, which now is a collector's item.

Model 720

The John Deere Model 720 is one of the most successful two-cylinder tractors. It was a direct replacement for the already popular Model 70 and was promoted as a five-plow tractor. On many farms, it could replace two

Continued on page 89

Michael (Mike) Mack of Waterloo, Iowa, is the retired long-time director of the John Deere Product Engineering Center in Waterloo. Mack first worked as a young engineer at the Dubuque factory. In 1956, he moved to the newly opened product engineering center in Waterloo. At the peak of production, he estimated that the Waterloo facilities employed about 16,000 people for Deere & Company. He shares these memories of the two-cylinder era.

When did engineers begin using computers?

"Probably nothing had a more significant impact to tractors, to tractor engineering, to the entire industry, in all the years I was associated, than the advent of the computer. It started to happen about 1956, at the end of the two-cylinder era. The first pair of gears that were designed by a computer were my gears.

"I did the calculations for the final drive and spur gears for the high-crop derivative of the Model M. That was the first tractor where I was the project head. The tractor was up in the air. For longitudinal stability, we extended the Model M chassis maybe seven inches so it wouldn't tip over backwards. I had done the manual calculation for the gears. It took a fellow who was fairly practiced several days to do the calculations by the time you checked everything. The dimensional features had to be carried out to eight decimal places. We had to check every single decimal place for tooth thickness, backlash, surface compressive stress, beam stress, other stress, etc. You couldn't get enough decimal places on a slide rule. Those were all done with a manual calculator.

"A good friend of mine, Harlan Van Gerpen, wrote a computer program for the same set of gears. He took the program to Moline and ran it on their old IBM 650, which filled the equivalent of almost a whole room. Everything was done with cards and it was scheduled pretty closely. He ran the program during the night. He came back the next morning with the data and he put it alongside of my manual data, which was just a handwritten sheet. When it came back, every single item of data to the eighth decimal place from his printout sheet to my manual sheet checked exactly.

"The place was pandemonium with the success of this program. The computer took about 15 minutes to run this thing and I had spent days on the manual calculations. Today that sort of calculation would be completed in less than 10 seconds. That was the beginning of the design of gears on a computer. Everybody worked on a drawing board. Today nobody works on a drawing board. That was way before the days of computer graphics. It was a very, very exciting moment; probably one of the most exciting moments I can recall."

What drove the product changes?

"There was a coordinated effort between what happens out in the field and what the farmers wanted. Of course, concepts developed at the engineering level are very interesting and fascinating, but you have to be cautious about it. The engineering manager has to be careful that he doesn't let some persuasive young engineer with an exciting concept get going on something that is fun for the engineer, but maybe there's not a market for it. You have to balance what the trade wants in the field with what engineers can actually do in the laboratory. Truly new concepts are usually developed at the laboratory level, but there's no use spending a lot of engineering dollars until you know they have practical applications.

"Deere had a practice at that time when we were going into the field to send a team. Maybe six or eight of us would go into the field every year. We would scatter to different directions and make visitations to lots of dealerships and talk to them about what they liked and what they didn't like. We would gather this product information and lots of times we would solicit ideas. It didn't necessarily mean we would follow through, but we would get the information. If they had some pet peeve about a product, we were most interested in hearing that. They had lived with the product in the field and they had a feel for it. We listened carefully to them and usually made corrections."

What made Deere different?

"Well, you can get different answers. I ended up starting to observe the competition. I think the concept of having Deere work close to the dealerships and solicit their input was important. They did more of that than most competitors, and that was a powerful, powerful bit of information.

"Then, of course, Deere had a very elaborate and sophisticated test program at the engineering center. We didn't usually get into production without making sure the marketing people were in step. They had to be positive about it. At the same time, you had to be sure you built sufficient prototype samples. You'd look at the level of failure and you'd make projections. You might go with ten prototypes to get a feel for the failure rate, but then you'd make some pretty sophisticated statistical analysis of what would happen when you started making 100 a day. I think Deere's concept along those lines was a step ahead of competitors.

"If you look at Deere's success, there are some things they must have done very right. Deere's factories, for the most part, were in relatively small communities. When you've got a huge factory in a fairly small community, there's not a lot of turnover. A very high percentage of people who started working for Deere as young men would retire there. Something inherent with Deere is that there isn't a lot of turnover at the factory level. In a large city, they'd walk out the door of one factory and into the door of another factory. The same skills are required. You had continuity over the years at Deere, and I think that has a lot to do with Deere's overall success. It was expressed in the quality of the product. If people did the same job for many years, they inherently became very skilled at it.

"Factories for tractors and cars bought steel from the same companies. The steel they started with wasn't all that different. The same was true for the machine tools. At the awards banquets and retirement banquets, we made this point: our competitors don't have our people. It's true. You didn't see our people walking out of one factory door and into another.

"Deere was very fortunate in having leadership that was insightful. They didn't make too many bad guesses. There were two factories. Although they had certain similarities, they had different factory managers and different chief engineers. They had a delicate balance, being loyal to your company, and yet feeling they had some freedom to stake out their own ideas. That's what we witnessed in those years."

Continued from page 87

tractors for chores like plowing and disking, and it handled as easily as most small tractors.

The first Model 720 was built on July 12, 1956, two days after the first Model 820 was built at Waterloo. The last 720 was completed on July 28, 1958. During its two-year run, Deere built 30,675 tractors in the Model 720 series. Buyers had options for gasoline, All-Fuel, LP-gas, or diesel fuel engines, and options for four configurations: row crop dual wheel, row crop wide front, standard tread, and high crop. The engines had major improvements, such as a new alloy steel crankshaft, heavier connecting rods, larger piston pins, improved cyclonic

swirl in the combustion chamber, and higher rpm. Engineers also improved the pistons, cylinder head, and ignition system.

A Model 720 equipped with a gasoline engine was the largest two-cylinder row crop tractor that Deere & Company tested at Nebraska. It set numerous records, such as 56.84 belt horsepower with gasoline, and set a new fuel economy record. The engine operating speed was 1,125 rpm. Previous engines in the Model G and Model 70 tractors had turned at 975 rpm.

The Model 720 was a heavier tractor and hit the scale at 6,800 pounds—about 660 pounds heavier than its big and beefy Model 70 predecessor. A 1,000 rpm

This Model 720 has the classic green dash and metal steering wheel. The six-speed transmission was standard, but this tractor has the optional high-speed fourth and fifth gears.

PTO was optional, although it was seldom ordered. The Model 720 had a six-speed transmission that replaced the outdated high-low range Model G transmission. Although the Model 70 continued to feature the high-low transmission, the six-speed transmission offered an improved three-point hitch.

Models 70 and 720 had the same specs for the diesel engine bore and stroke. The diesel came with the V-4 pony engine that used two levers rather than one. One lever decompressed the engine and the other engaged the pony pinion to the flywheel ring gear. A 24-volt electric starting system was added as an option in February 1958. Late in the run, the big diesel was given an anti-reverse camshaft with better lobe profiles. It would prevent the possibility that the diesel engine could start to run backwards when it was shutting down.

The four engines went through the Nebraska tests around the same time. The LP-gas recorded 52 drawbar

Dual remotes and a three-point hitch decorate the back of the Black Dash Model 720 diesel standard. It also has the V-4 pony motor, an adjustable front end, a full set of wheel weights, external air stack, and pre-cleaner.

This 1958 standard-tread Model 720 diesel is ready for a ride down memory lane. Knabb says it arrived "literally in a basket" and with two pickup loads of parts. This was his first diesel restoration project, and it took about 30 months of work.

This close-up shows the hydraulic hoses that connect the dual remotes to the front-mounted rockshafts. This is a rare feature for any 20 Series tractor and enables the operator to raise each side of the cultivator independently.

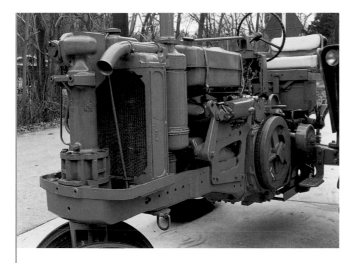

This is an electric-start version of the 1958 Model 720 diesel. This tractor is nearly through its restoration and is waiting for the hood and other sheet metal to arrive with new paint. Approximately 2,000 Model 720 tractors received an electric start in 1958. This tractor also has the rare front-mounted rockshafts.

and 57 belt horsepower. The gasoline engine recorded 53 drawbar and 57 belt horsepower. The diesel recorded 51 drawbar and 56 belt horsepower. The tractor fuel (All-Fuel) version recorded the lowest results of 40 drawbar and 44 belt horsepower.

The basic Model 720 with its split pedestal was by far the most popular version; nearly 23,000 models were built. Deere built 4,500 of the Model 720 Standard tractors. Nearly 3,600 of these were diesel. Only 80 had the All-Fuel option and most were exported. The 720S could be ordered with either a 55.5-inch fixed-tread front axle or an adjustable axle with 52- to 58-inch tread.

Model 730

The John Deere Model 730 two-cylinder row crop tractor was the epitome of the two-cylinder tractor era. It had all the bells and whistles of two-cylinder craftsmanship. It was one step down from the largest power unit,

This 1959 John Deere Model 730 is owned by Dave Jones of West Chester, Pennsylvania. It's one of only 292 that were built with gasoline engines.

the 9,875-pound diesel Model 830, but it was designed for row crop work and was more versatile on the farm. Where the 830 had a singular purpose to provide power to perform the heaviest drawbar and belt work on expansive fields of grain, the 730 could do nearly all of that while providing the steering accuracy and power assistance required for row crops. It was a thoroughly modern, highly capable tractor. Many of these models are still at work today.

The 30 Series was introduced in 1958, just two years after major upgrades in developing the 20 Series. By 1958, Deere & Company was well along in gearing up for its New Generation tractors with four-cylinder engines. The first of these, the mighty 8010, was introduced in September 1959. Rather than invest in major alterations at this point, Deere marketing focused on operator comfort and modern styling. The plan worked.

The series sold very well and has remained highly valued by owners.

A 30 Series tractor was easy to spot. The painting scheme was quite different from a 20 Series tractor. A simple change had made it cleaner and much more attractive. On the driver's seat, operators saw more. These tractors came with a new sloping, automobile-style steering wheel and with a sloping dash where instruments were clustered for easy viewing. The new row crops series came with front mounted headlights. This headlight feature was highly practical, as well as an important styling innovation.

The Model 730 was a mechanical twin to the 720. Configurations offered included the row crop, the standard, and the Hi-Crop. All three came with a six-speed transmission and four engine choices: gasoline, All-Fuel, LP-gas, or diesel. Rubber tires were standard, although it still was possible to order steel front or rear wheels. Basic features on the Model 730 included power steering, an adjustable front axle, three-point hitch, Float-Ride seat, pre-cleaner and air stack, dish-type rear wheels, and dual hydraulics. The 730 Diesel had options for the V-4 starting engine or a 24-volt electric starter. The 730 had styled flat-top rear fenders with an improved lighting system for work at night. For operator safety, the fenders had handholds that protected the operator from mud and dust and from possible contact with the tires. A convenient step in front of the rear axle made mounting and dismounting both safer and easier.

At Waterloo, the Model 730 was the end of the line for two-cylinder tractors. The first Model 730 was built at Waterloo on August 4, 1958. The last tractor on the Waterloo line was built on March 1, 1960. The two-cylinder assembly lines in Waterloo had seen more than 40 years of production. However, soon after, the 730 assembly line and production tooling was packed onto a boat and shipped to a new Deere & Company tractor facility in Rosario, Argentina. New Model 730 tractors were manufactured at the Argentina assembly line until 1970. More than 20,000 Model 730 tractors were built there. Deere also assembled other Model 730s at a plant in Monterrey, Mexico. Before the 1961 shutdown, Waterloo built nearly 30,500 units of the 730 for the North American market and almost all were row crops. About 3,200 were in the standard configuration and about 230 were Hi-Crop models.

Model H

Late in 1937, Deere & Company engineers at Waterloo began drawing plans for their smallest general purpose row crop tractor, a scaled-down form of the Model B. The two successful row crop tractors, Model A and Model B, both were being scaled up in size and power. The unstyled Model B had started as a one-plow tractor and, as of 1939, would belong in the two-plow tractor class. It would be the lowest-priced Waterloo tractor, but it had to retain the John Deere rugged construction and reliability. With the technical know-how already in place at Waterloo and with little need for field trials, the factory tooled up for the new Model H in a very short time. Less than a year from the first drawing

This John Deere Model 720 row crop, gas-fired tractor is one of several versions of the Model 720 that is owned and was restored by Ralph Knabb of Spring City, Pennsylvania. This tractor was shipped on December 20, 1956, to Omaha. Around 2004, Knabb completely rebuilt everything except the transmission and finished the restoration as close as possible to factory specs.

This 1945 John Deere Model H tractor is owned by Roger and David Dills of Brockport, New York. Production of the smallest row crop tractor from Waterloo lasted for only eight years. It was running when purchased in 2000, but has undergone a thorough restoration.

stages, on October 29, 1938, Waterloo tagged its first Model H as built and ready for sale.

The Model H, successor to the Waterloo heritage of Waterloo Boy, A, B, C, D, and G, was produced as a styled tractor from the beginning. It looked like a small Model B with the tricycle wheel configuration. At the Nebraska test, it weighed 3,035 pounds, which was 850 pounds less than the Model B. It could do two-row work on most fields. At $600, it was affordable for the market niche. As it turned out, Model H was well suited to a new market. Farmers already equipped with a Model D or a large row crop tractor liked the Model H for its ability to perform light jobs at low cost.

At dealerships, the Model H became the filler tractor between the row crop tractors and the even smaller Model L. Waterloo built approximately 60,000 Model H tractors, including three variations before the end of the run on February 6, 1947. Due to wartime shortages, production was shut down from May 1942 to April 1943.

The Model H was and remains a neat little tractor as long as you're not in a hurry. It was anything but nimble and quick. On good roads and using a governor override, drivers could kick up the engine speed to 1,800 rpm to achieve a top speed of 7.5 mph. It traveled on 32-inch rubber tires. When it was time to slow down or stop, the brakes applied directly to the axle. The rear tread was adjustable from 44 to 84 inches.

Waterloo engineers put the smallest two-cylinder horizontal engine ever built by Deere & Company under the hood of the Model H. The cylinders were only displacing about 100 cubic inches with each rpm, but they raced along at 1,400 rpm, which was faster than any other Waterloo two-cylinder engine and second only to the Moline-built Model L. The Model H belt pulley drove off the camshaft rather than the crankshaft. This change allowed engineers to increase the engine speed. It also meant the belt pulley operated counterclockwise, reverse to the standard direction.

Model H, like the Model B, burned distillate. The new Model L engine from Moline fired on higher-octane gasoline. All three were taken to the Nebraska Tractor Test station between September 6 and November 4, 1938. On the Nebraska test, Model H kicked out a little less than 12.5 horsepower on the drawbar and nearly 15 horsepower on the belt, placing it midway between power ratings for Model B and Model L. Maximum pull at Nebraska was 1,839 pounds, which was slightly more than the original Model B had produced four years earlier. The little engine also set a fuel economy record.

According to historian J. R. Hobbs, the Model H "was perhaps the most versatile and adaptable tractor in its class despite the claims made by Ford-Ferguson for its 9N. Even more versatility was made possible by ordering a PTO, priced at a most reasonable $18."[11]

[11] The First Numbered Series of John Deere Tractors: Models 40, 50, 60, 70, and 80. Hain Publishing, 2004, page 70

Here is another view of the Model H from the shadow side. It's a small tractor and easy to operate. It's very useful for light work, from pulling wagons to single-bottom plows.

Production of Model H went into high speed in mid-January 1939 and it soon was outselling the popular Model A from Waterloo. There were three significant areas of improvements in the 1939 models. These changes were to the camshaft and all components on each end of the shaft, the shifter section, and the slider shaft with its associated components. There also were changes in the brake latch, drawbar, brake shoes, cast front wheels, rear wheel rims, flywheels, steering wheel, and more.

New features were added in 1941, including the hydraulic power lift. It was driven from the engine governor shaft to provide live hydraulic power to operate one or two hydraulic cylinders. An electric starter and electric lights were added in mid-1941 at serial number 27000.

Farmers liked the narrow, tapered hood. Operators had great visibility for the work ahead. Many implements were developed to work with the Model H, including plows, tillers, cultivators, planters, hoe, rake, binder, harvester, mower, and more.

Waterloo offered only three variations to the basic Model H, as compared to six or seven variations for the larger row crop tractors built at this factory. The first variation, Model HN, was introduced in April 1940. The HN had a single front wheel. It was excellent for narrow row crops. In over seven years, Waterloo built nearly 1,000 HN tractors.

The second variation is among the most rare of all Waterloo tractors. Model HNH had the HN single front wheel but the rear axle held 6-inch-taller wheels that gave it a nose down to the ground appearance. Waterloo only built 37 of this model.

The third variation, Model HWH, was a group of high-and-wide tractors for vegetable growers. Only about 125 HWH tractors were built between February 1941 and March 1942. They became known as the California Hi-Crop H. The chassis had six inches more ground clearance than Model H. The axles could be adjusted to 68 inches wide on the front and to 84 inches wide on the back.

Due to wartime shortages, Model H production was shut down from May 1942 to April 1943. Waterloo built a total of approximately 60,000 Model H tractors before the end of the run on February 6, 1947. Model H had been the final two-cylinder era tractor family introduced at the Waterloo Works. It was the smallest of the styled Waterloo-built row crop tractors. It also had the shortest time on the assembly line of eight years and three months.

A rare 1955 John Deere Model 40 Hi-Crop tractor is hiding in the corn. It was purchased and restored by Phil Pokorny of East Aurora, New York. This is No. 158 of the 294 Model 40 Hi-Crops that were manufactured.

CHAPTER
6

DUBUQUE FACTORY
FIRST COUSIN TRACTORS
MODELS M, 40, 320, 420, 330, 430, 435, 440

The Dubuque Tractor Works' first tractor was a Model M, completed on March 12, 1947. A month later the Model M began arriving at dealerships, as a replacement for the Model H tractors they had been receiving from Waterloo. The improved Model M was followed by several more two-cylinder tractor families from Dubuque. The Dubuque tractors were scaled down in comparison to the remaining three row-crop tractor families and heavy duty standard-tread family built in Waterloo.

Structurally, there was a big difference between Waterloo and Dubuque tractors. The Dubuque tractors had less casting material and weighed less. The basic Waterloo tractor was composed of a few large casting pieces. The Dubuque tractor, disassembled, was unrecognizable as a loose collection of bolts and pipes and brackets. In the field, it often needed front wheel weights to hold it down.

Dubuque took a different approach to the two-cylinder engine. The pistons and valves were vertical as opposed to the horizontal arrangement for pistons and valves manufactured for Waterloo engines. Overall, Dubuque tractors earned an excellent reputation for reliability. People with long experience in restoration say that Dubuque tractors were well built with strong components. The components, as a rule, had a very long life expectancy.

Dubuque built a tractor transmission that was engaged with a modern foot clutch. The Waterloo tractors all used a hand clutch. Today, learning to operate a hand clutch safely can be a challenge for a younger collector. To engage a hand clutch while backing up, the driver pushes his arm forward while looking backwards.

Dubuque developed the "Liquid Brain" or Touch-O-Matic as a precise implement positioning system, equivalent to Waterloo's Powr-Trol system. It was introduced with the 1947 Model M. Engineers provided the live hydraulic power by mounting a hydraulic pump on the front of the engine. The operator could have single Touch-O-Matic on the Model M or dual Touch-O-Matic on the Model MI or MT. Dual operated each lift arm independently. Touch-O-Matic worked very well and was kept in production as an option through to 1960.

Dubuque also developed a new attachment system for implements. Most Dubuque tractors had the Quick-Tatch system. With this system, an operator could back into a waiting implement, put in one or two pins, and drive away with it. Tillage implements could be leveled with a special yoke under the tractor.

This 1948 John Deere Model M was one of the first tractor models built at the Dubuque, Iowa, factory. This tractor's current owner, Paul Lesher of Wernersville, Pennsylvania, purchased it from the original owner and restored the Model M in 2004.

Model M

March 12, 1947, was a big day in history. In Dubuque, Iowa, the new John Deere factory tagged its first Model M tractor as built. Bigger changes were coming. In the Middle East, British military units captured most of 800 Jewish settlers after their boat was beached north of Gaza. It was the beginning of the Exodus and the birth of modern Israel. In Washington, D.C., at a joint session of Congress, President Harry S. Truman introduced his plan to fight advancing communism. It was the start of the Cold War.

Between 1947 and September 8, 1952, the Dubuque Tractor Works went on to build 87,000 tractors and industrial crawlers in the Model M family. After that, the Model M was sturdy enough to stay working on farms and job sites for the next 50 years.

The Model M was in development during the war years. In 1945, the mission of the new Dubuque factory was to manufacture a new small tractor, the Model M, to replace the L and LA tractors that were manufactured in Moline. The M was to be a simple, versatile tractor for small farms. Like the Model H at Waterloo, it would be smaller than the Model B family, but larger than the little Model L/LA family. Dealerships could promote it as a tractor that was slightly larger than the

Covered with a blanket of fresh snow, this original Model M tractor is looking pretty good after six decades.

This 1950 industrial version of the Model M tractor, the MI, is owned by William Spiller of Broomall, Pennsylvania. The Model MI was a small, standard-tread tractor and was often used for work by highway and street crews. Only 1,034 MI tractors were built, and nearly all were painted yellow. This rare item was painted orange for use at an airport in Iowa.

This detail of the Model MI engine shows that it is in better-than-showroom condition.

Model H, with the latest bells and whistles of 1940s technology and a completely new engine. Its competitor was the super-successful Ford-Ferguson 9N that had been introduced in 1941.

The Model M went through the Nebraska Tractor Test on October 5, 1947, and was the smallest and last of five John Deere tractors on test that year. The Model M engine, spinning at 1,650 rpm and burning gasoline, generated a maximum of 18.1 horsepower on the drawbar and nearly 20.5 horsepower on the belt with virtually the same displacement as Waterloo's Model H engine from nine years earlier. It was 40 to 50 percent more powerful and could pull a load of 2,329 pounds, which was about 500 pounds more than the Model H. It also was heavier by 900 pounds. Model M boasted an improved four-speed transmission. A three-point hitch was never offered on this model.

The standard-tread Model M was promoted as a one-row tractor. Standard equipment included Touch-O-Matic, PTO and electric starting. It boasted an improved four-speed transmission. A distillate-burning engine also was available. There were options for an adjustable front axle, a belt pulley, and larger tires. Dubuque built about 45,000 of the Model M tractors between 1947 and 1952.

The Model MT went into production in December 1948. All the MT tractors were taller and more versatile for row crops and vegetables than Model M. The Model MT offered rear wheels that were fully adjustable on the axle from 48 to 96 inches wide, at any setting the operator chose, and one man could do the job of resetting the wheels. It also had the first Dual Touch-O-Matic, giving independent control of front and rear-mounted implements.

It was offered with a choice of three front-ends: a single front wheel, a dual-wheel tricycle-type, and a wide adjustable type. Among the three MT front-ends, the dual-wheel tricycle was by far the most popular. The N version of Model MT with a single nose wheel was chosen by a few growers who wanted it for specialized vegetable production. It could do a clean job of cultivating in narrow-spaced rows without damage to the plants. The MT front-end with a wide front probably was the most rare. The MT W version could straddle raised beds. Dubuque built about 30,500 of the MT tractors.

Two versions of the Model M are recorded in *John Deere Tractors & Equipment, Vol. 1*. The standard-tread

This 1951 John Deere Model MT is as green as the corn it is cultivating. The tractor is owned by Steve Witham of Cherry Valley, New York.

Model M was built to cultivate one row. Unlike the Model H, it had fixed, matching wheel spacing, front and back. Model MT was a popular, taller, tricycle version with a single front wheel and wide rear axle that was built from 1949 to 1952. It could cultivate two rows and had a supply of two-row equipment. Dubuque built about 30,500 of the MT tractors. A few MT tractors had the height of the H, but a wide front end was similar to a Hi-Crop tractor.

The Model MI, a yellow industrial version, was built between November 2, 1949, and October 8, 1952. Dubuque produced 1,032 of these tractors. Compared to the Model M, it had a new drawbar/hitch design with a mounted toolbox, flatter fenders, and a final drive housing that was rotated 90 degrees. The whole tractor was lower than Model M and it had a lot of options. It was usually highway yellow, although some were orange. For New York state, the tractors were painted in blue and yellow.

The Model MC, a crawler version with three track rollers, was built in the same period and was very popular.

A shiny new Touch-O-Matic decal highlights a transmission that was innovative when it was introduced on the Dubuque tractors. The Model MT had dual Touch-O-Matic, while Model M tractors had single Touch-O-Matic. This Model MT is owned by Steve Witham.

This MT-200 cultivator was made for the MT tractor and could handle two rows. A full line of matched working equipment was offered for the MT tractors, which made it able to handle nearly any task on the farm.

It was the first high production crawler from Deere & Company. Dubuque built about 10,500 of these crawlers. Essentially, the MC was a set of tracks on the chassis of a Model M. It was used on small farms, in construction, and in forestry. Some were industrial yellow, but most were John Deere green. It was possible to build an orchard version of the Model M. It wasn't created at the factory, but at least one dealership modified a few in the orchard configuration. At least five of these modified tractors are known to exist.

Retired chief engineer Mike Mack recalls this anecdote from the days the Model M was being built.

"There was an oil filter can on the side of the engine. The oil filter screwed onto the can, just like it does on most engines. It had a very deep draw. It was probably six or eight inches long and the can was probably three or four inches in diameter. There was a spherical surface at one end. The other end was open. Deep draw. They had to shut the whole factory down because they couldn't ship tractors out without filters. The thing went on for days

and days and days; tractors kept accumulating. Finally, I believe they might have shut the factory down for a while.

"They tried everything. They called in die lubricant experts from Detroit who made a living designing dies for large presses. Nobody knew what the problem was. They tried all kinds of die lubricants; no luck. I assume something was out of spec with the material with the sheet steel. When the sheet steel was brought into the factory to be cut up into pieces for the draw, it was called 'pickled in oil.' It had an oily surface.

"A young man that I knew very well at the time had the idea of sprinkling some Duz (soap) on one of the sheets and put it through the press. It was like the day that Columbus discovered America! Everybody was so happy and celebrating they'd found a solution. That was the answer, at least to get by temporarily.

"I remember going out in the shop ten days later and there were many, many large cartons of Duz that they were using. It worked. It put them back in production. I

continued on page 104

THE DUBUQUE FACTORY

On December 31, 1944, Deere & Company announced that it planned to build a 600,000-square-foot factory in Dubuque, Iowa. The factory's initial purpose would be to make shell casings for a government contract during the war. When World War II was over, the factory would build tractors. It was the first tractor factory built from the ground up by Deere & Company. The estimated cost was $9.4 million. Dubuque had a small population of about 45,000 people, but it was centrally located between Waterloo Tractor Works and the company's general headquarters in Moline, Illinois. The Dubuque area

offered good labor supply, good water supply, shipping by rail or truck, and an option for shipping on the Mississippi River. The site selected was roughly 740 acres and about three miles north of Dubuque in the Peru township. Maurice Fraher was appointed general manager in December 1944.

Grading began on June 25, 1945. Company officials envisioned a modern, one-level factory rather than the multi-level pre-war building style. Deere & Company Vice President L. A. Murphy wrote: "This is unquestionably the largest single undertaking Deere & Company has ever attempted, and the problems connected with the production of an entirely new product in a brand new plant in a strange town is a tremendous task..."[6]

Original plans called for 19 buildings, but before these were completed it was decided to expand the factory by 50 percent. The buildings included three machine shops; a foundry for manufacturing tractor and engine parts; a shop for heat treating, forging, and stamping sheet metal; an assembly building; a powerhouse; and many supporting buildings for a comprehensive manufacturing facility. The name changed three times in the two-cylinder era. The first name was John Deere Dubuque Tractor Company. By October 31, 1946, the new factory had 661 employees.

During the start-up of the new factory, many tools, fixtures, patterns, and dies that had been used on the Model L and Model LA tractors were transferred from Moline to Dubuque. The first engine came off the line in September 1946 and was shipped for use in a combine at the John Deere Harvester Works in East Moline. In November, the Dubuque newspaper reported that 40 combine engines were coming off the line each day. Tractors would begin assembly as soon as the required foundry machine could be obtained and installed.

The first Model M tractor was completed on March 12, 1947, according to the *Dubuque Telegraph-Herald* and other records. A run of 50 tractors was completed on March 28. On April 1, the first carload of 10 tractors was shipped, including the first Model M. It was shipped to the Arizona farm of Deere & Company president, Charles Deere Wiman. In August, the factory's name changed to John Deere Dubuque Tractor Works of Deere Manufacturing Company. By October 31, it had 1,754 employees with an hourly pay rate between $.96 and $1.61. It had two unions, the FE-CIO and the Pattern Makers League of North America. It also turned out replacement parts for small gasoline engines, which was a job formerly done at Moline and Waterloo.

In 1949, the lines expanded to handle construction, forestry, and agricultural equipment. The MC crawler was introduced and became very popular in forestry and housing construction. Many

consider it the first true John Deere product used for industrial applications. The crawler was followed by the MI series of industrial wheel tractors in 1950.

In 1952, James Wormley was appointed general manager. The fiscal year ended on October 31 with 1,263 employees. The mid- to late 1950s began a new focus for the plant with its interest toward providing more industrial tractors with allied equipment. By 1955, hourly pay had risen to $1.55 to $2.99. In 1956, Lloyd H. Bundy was appointed as the third general manager. The John Deere Industrial Division was created this year, and followed by establishment of an industrial marketing division, industrial equipment dealership, and an industrial engineering department in 1957. On August 1, 1958, the name changed to John Deere Tractor Works of Deere & Company. The Works now employed 2,219 workers.

As the two-cylinder era ended, the Dubuque factory introduced multi-cylinder integral bore engines designed for use in combines and power units. In the 1960s, the Dubuque Works entered a time of steady growth with agricultural and industrial tractors: the Models 1010 and 2010. In 1960, the plant had 3,800 workers on October 31 and engineers had started work on a new diesel engine.

A Model 440 industrial crawler was the last two-cylinder tractor off the assembly line at Dubuque on March 31, 1960. In 13 years of tractor production, the Dubuque Works had built nearly 227,000 two-cylinder tractors in about 70 model variations.

Fifty years later, the John Deere Dubuque Works employed about 2,350 workers and was planning a new 16,000-square-foot multi-purpose facility with a reception area, 200-seat auditorium, factory store, and product display area. Today, the site covers 1,443 acres and 4 million square feet. It is responsible for designing and manufacturing world-class construction and forestry equipment.

[6] *John Deere Dubuque Works 1947-1997: Changing Perspectives,* by Faith Hamilton Meyer, published 1997 by John Deere Dubuque Works, page 4

Continued from page 101

suppose, eventually, they did something to the steel, but that's a long process, to make changes in the steel. It was a very exciting moment. There were large, large quantities of soap boxes in the factory, and the popular slogan around the factory was the old Duz commercial, 'Duz does everything'."

Model 40

Deere & Company replaced the popular Model M series with a numbered version, the Model 40. It was the end of October 1952. Across the Pacific Ocean, the Korean War was fought. The next day, on October 31, the United States detonated its first hydrogen bomb at the Eniwetok Atoll, with a 10.4 megaton yield.

In March and July 1952, Deere & Company had replaced two row crop tractor family names (Models B and A) with Number Series designations, Models 50 and 60. The Model 40, from Dubuque, was introduced as a 1953 model. Priorities achieved in the second Dubuque factory tractor included the first true three-point hitch on a John Deere tractor, improved operator comfort and convenience, more horsepower, good fuel economy, and more.

The Model 40 was styled with paint and sheet metal to match Models 50 and 60 at the dealerships. It had more space on the operator's platform than Model M tractors, a deeply cushioned seat, new three-point hitch with an exclusive load and depth control, Touch-O-Matic hydraulics, and a wide range of implement options. Four months into production, buyers were given the option of an All-Fuel engine for an extra $36.

The Model 40 retained the Model M gasoline engine, but this little tractor had 25 percent more horsepower than the M series. Compression was increased to 6.5:1. The rpm was boosted to 1,850. It tested for 22.9 horsepower on the drawbar and 25.2 horsepower on the

A rare 1955 John Deere Model 40 Hi-Crop tractor is hiding in the corn. It was purchased and restored by Phil Pokorny of East Aurora, New York. This is No. 158 of the 294 Model 40 Hi-Crops that were manufactured.

brake. Maximum pull was 3,022 pounds with a 700-pound increase. Weight also went up by about 600 pounds.

The factory built about 49,000 Model 40 tractors by the end of the last production run in November 1955. There were seven versions, which were enough to justify a small collection, and changes happened during the production run. For instance, there was a steering wheel change. The later steering wheel was larger to make it easier to steer when equipped with a loader or other front-end attachment. There are three steering gear assemblies. The later steering sector has a second pin to provide more wearing surface.

Top three in the Model 40 family are the tricycle, utility, and crawler versions. More than 17,000 Model 40T tricycle-type tractors were built. They had the gas engine, except for about 470 that burned All-Fuel. Another 11,000 models with gasoline engines were built in the 40S

utility version and a nearly equal number were built as the 40C, the crawler editions. Respectively, the All-Fuel engine was supplied for 650 and 280 in these versions.

Less popular variations included the 40 Hi-Crop, 40 Standard, 40 Special, and the 40W, two-row Utility. Here's how they were promoted at the time in John Deere literature.[12]

40 Standard: The two-plow general purpose tractor for speedy, economical work on all tillage jobs, mowing, hauling, and one-row planting and cultivating in corn, cotton, tobacco, and similar crops.

40 Tricycle: The two-plow general purpose tractor for all-around farm work that furnishes two-row capacity in corn, cotton, tobacco, and four- or six-row capacity in beans, beets, lettuce, and similar crops.

40 Hi-Crop: The two-plow tractor with extra-high clearance (32 inches under the tractor) and wide wheel

The Model 40 Hi-Crop had 32 1/2 inches of ground clearance, adjustable wheel spacing from 54 to 84 inches both front and back, and a turning radius of 10 feet plus half an inch.

[12] The First Numbered Series of John Deere Tractors: Models 40, 50, 60, 70, and 80, Green Magazine, 2004, page 19, by Hain Publishing.

This pristine 1956 John Deere Model 320 with standard tread heads up the 20 Series collection of tractors owned by Ralph Knabb of Spring City, Pennsylvania. This is one of the most prized John Deere two-cylinder tractors. It was the smallest of the Dubuque tractors, was built in low numbers, and is extremely hard to find today.

spacing for cultivating tall, bushy, or high-bedded crops. The 40 Special Model, not shown, provides 26-1.4 inches clearance.

40 with Single Front Wheel: This model is the same as the 40 Tricycle except for the single front wheel which provides extra clearance in narrow-spaced rows. A favorite with vegetable growers everywhere.

40 with Wide Front Axle: This is another variation of the tricycle model for growers who wanted an extra-wide wheel adjustment on the front axle as well as the rear.

40 Crawler: This rugged, highly maneuverable track-type tractor is used on many farms throughout the world, and it is a favorite for light logging and contracting jobs. Pulls three-plow bottoms. Available with either 4- or 5-roller track frame.

40 Two-Row Utility: It retains the low, stable, overall design of the 40 Utility, but has increased crop clearance

and wider wheel treads to straddle and cultivate two rows of corn, cotton, beans, peanuts, and similar crops.

Model 320

An entirely new family of tractors was introduced by Deere & Company in 1956 as the Model 320 tractors. On July 25 of that year, Mickey Mantle was on the cover of *Time* magazine, along with mention of a southern politician, Lyndon Johnson. Sadly, off the coast of Cape Cod, a horrific ocean collision took the lives of 51 people when the Swedish liner *Stockholm* rammed and sank the Italian liner *Andrea Doria*.

The Deere & Company 20 Series tractors had been introduced in late 1955 by the Dubuque division. Generally the 20 Series tractors offered a 20 percent power increase. The Model 320 moved into the long line at the low-power end as a tractor for small farms and as a backup for light work on larger farms. Dubuque

This rear view of a Model 320 shows the unusual belt pulley arrangement, which was located below the driver's seat. The pulley was powered by the PTO.

patterned it after the Model M. The 320 weighed about the same as the Model M or Model 40. It had the M engine and was priced a little higher than the Model 40, but less than the Model 420 (introduced in late 1955). In addition to the basic tractor, the Model 320 was offered in variations targeted for vegetable growers and for very small farms.

The Model 320 series found its niche mostly in the South, but it was never a big seller. It must have been a disappointment for the company in sales volume. However, low sales and small size combine to put it at the peak of the price list for today's discriminating tractor collectors.

Complete production of the Model 320 was approximately 3,080 tractors. They were built in two distinct phases. The last was built on July 30, 1958. Approximately 2,560 Model 320 tractors were built in what became known as Phase One. For these, the steering wheel was straight up and down relative to the operator. For Phase Two, which came late in the production

Here is a side profile of a perfectly restored Model 320. These tractors had only a few options. They were lightweight, quick, and versatile. The 1958 list price was about $1,900, which was only $76 less than the Model 420 standard. Today's value for a restored Model 320 is around $15,000.

as the 30 Series was being prepared, the Model 320 came out with a slanted dash and slanted steering wheel. Approximately 520 tractors were built and these new features made them much more comfortable for operators.

About 60 percent of Model 320 production was in the basic model, but there were two variations and some other differences during the production run. The Model 320S (standard) had a 21-inch ground clearance. It appealed to growers of peanuts, vegetables, berries, and tobacco. The 320 Southern Special was a variation of the 320S and was built for a handful of farms in Louisiana and Texas. The 60 to 70 Southern Specials had even higher clearance for work in certain vegetable crops.

The 320U (utility) had a lower stance than the basic Model 320. It was shortened five inches by a combination of shorter spindles in the front and geared, offset rear axles (rather than direct drive axles). Its special appeal was for orchard operators, people in the mowing business, and those working in confined areas. Deere built about half as many of the little 320 utility tractors as it did standards. A few Model 320 tractors were purchased for industrial use, such as highway work, and were painted bright orange or yellow.

The Model 320 had the same engine as the Model 40, with a bore and stroke of 4x4 inches. It was never tested in Nebraska. Another very small 320 group, less than 20 tractors, had the All-Fuel engine.

Popular production features were retained on the 320. These included disk brakes, push-button starting, the new Float-Ride seat that was adjustable for the operator's weight, Touch-O-Matic live hydraulics, and the load-compensating three-point hitch called Load-and-Depth Control. Early in production, the crankshaft and rods on the Model 320 were upgraded to those used on the Model 420. Sheet metal on the 20 Series was nearly the same as on the series it replaced. However, the 20 Series sported a new, flashy paint job with horizontal and vertical bands of bold yellow on the sides of the hood.

Model 420

The Model 420 was a tractor ahead of its time. It was the first of the John Deere tractors to be named by a three-number sequence. It came into production at the Dubuque factory on November 2, 1955, which was about eight months ahead of the other tractors that would become known as the 20 Series tractors. The earlier numbered tractor families became known as the early Number Series.

At that point in the mid-1950s, young Elvis Presley had a hit on the country music charts with "Mystery Train." November 2 was a historical day in the medical field because it marked the discovery of the polio virus vaccine.

Dubuque's Model 420 soon became one of the most popular small, row crop tractors of the 1950s. The Model 40 had been very popular, but the 420 was even better because it had significant improvements. In just two production years, Dubuque built about 47,000 of these little tractors.

Originally, Model 420 tractors were painted all green. At first glance, they looked identical to the Model M and the Model 40 they replaced. After Deere & Company shifted the painting scheme, in order to align with other tractor families in the 20 Series, the original Model 420 tractors became known as Phase One tractors. Phase Two Model 420 tractors had the new, flashy, yellow and green paint job. Making the most of the opportunity, Deere & Company offered a Direction Reverser option for all Model 420 tractors (except the Hi-Crop) as it introduced the Phase Two paint job.

Dubuque engineers gave the Model 420 a 20 percent power boost over the Model 40. They increased the cylinder bore a quarter inch to give the two cylinders a 113-cubic-inch displacement. A set of Model 420 tractors in three configurations was tested in Nebraska the following October. With the same rpm but more displacement than the Model 40, the little vertical, two-cylinder gasoline engine produced 29 horsepower on the belt and 3,790 pounds of pulling power. It had crept up into the power bracket of a Waterloo small-size row crop tractor, which was essentially equal to the Model 50 tested four years earlier.

Most Model 420 tractors were equipped with the gasoline engine, but about 5,000 burned either All-Fuel or LP-gas. Production of the All-Fuel or LP-gas units in

ABOVE: This rare 1957 John Deere Model 420 Hi-Crop cultivated flowers in an Alabama nursery for many years. It was purchased by Paul Champagne of Southampton, Massachusetts. After about two years of work, it was fully restored in the late 1990s.

some of the subgroups was less than ten tractors, which made them exceptionally rare and valuable.

Dubuque built eight configurations or versions of the Model 420 tractor on its assembly lines. Collectors can find five or six of the versions fairly easily, but two were low production lines. The 420C (crawler) was available in either four- or five-roller track configurations, like the 40C had been. With five rollers, it was longer and less maneuverable, but it had better flotation and stability. It was beefed up to handle chores in construction and forestry; it had an All-Fuel option as well. Many of the crawlers were equipped with a three-point hitch and ready to handle a cultivator on a hillside with lots of stability and pulling power. The pulling power of a 420 on tracks was 4,862 pounds, which was only 200 pounds less than the shipping weight. It was by far the most popular in the 420 family, with 17,800 built for the market.

continued on page 111

Ralph Knabb's 1958 Model 420 with the tricycle wheel configuration. This tractor has power steering, a slanted steering wheel, a five-speed transmission with continuous-running PTO, a Float-Ride seat, front and rear wheel weights, rear hydraulic outlets, a three-point hitch, and nearly every other option offered. The transmission was the only thing that didn't require complete rebuilding. The restoration was finished in 2006.

MEMORIES OF BILL HEWITT

"I worked as a student, from 1954 to 1956, graduated in 1957, and then went to work full-time. I worked in different departments of engineering and finally wound up in research. Everyone worked through a training program. For instance, when you were working in current design, you did machining tolerance changes to a cylinder head and you made changes all the way back to Waterloo Boy. That was the way you did it. It was done that way until Bill Hewitt came along. He was the guy that brought the company up into the twentieth century. He joined the company in 1954. I think everyone will attribute Bill Hewitt with bringing that company along and making it what it is today.

"When Bill Hewitt came to work at Deere, he sat on a tractor on his first visit and said something like, 'Fix that damn thing!' when he heard the two-cylinder. So they went away from building their muffler in the sheet-metal shop and pretty well got rid of, or at least certainly reduced, that pop-pop-pop sound. That was literally his first act on the factory premises. That was 1954, as I recall. I worked upstairs from the shop, but I can remember it echoing through the hall. The 30 Series had a new muffler.

"Hewitt had a great influence. He was quite decisive and quite a dashing figure. Everybody got to know him. He was quite personable. I remember many conversations with him and at that time, I was probably about as low as you could get in the engineering hierarchy—just a student starting out. I remember taking him on his first tractor ride on the final version New Generation tractors at a demonstration just before the introduction. They sent me out to drive the tractor when I came to work. I was a young kid. He was very highly respected by the company. He had the authority to put ideas into action. I understand he pretty well chaired the meetings quite democratically. After a little discussion, usually, they conceded to his demands and he got his way. There were a lot of pretty stout managers in the company, but he managed to put together many compromises they would accept and would act on.

"Hewitt would put people in charge and demand that they did get things done. He came out to the engineering center, got up in front of everybody, and said, 'We're going to make it or break it, fellows. I've bet the farm on it and you've got payless payday coming unless this sonofabitch runs.' That was the way he tackled a problem.

"About two years in advance of the shutdown to retool for the New Generation, Bill Hewitt personally charged Duke Kreitzberg, long-time manufacturing R&D manager, with purchasing all the new equipment and the tooling for the New Generation. Duke was given carte blanche to scour the world to find the most suitable equipment. Harley Waldon was the Waterloo factory manager; Bill Horan was the line manager on-site at the end of the assembly lines. [Bill had a counter in his office that counted each and every tractor off-line.] The factory line built unfinished two-cylinder tractors ahead of orders, preserved them, and then stored them everywhere in the Waterloo area. After that, the line was shut down for six months to retool [for the New Generation].

"Don't get yourself oversold. [The two-cylinder] was appropriate to the use. It was durable, it was adequate. In the research department in the 1960s, I did theoretical work with respect to the methodology of designing reliability. They kept records that probably don't exist today and I have read every single one of them; of the testing that was done clear back into the 1920s. It was documented very, very, carefully. I have read every single page of every single report up to the time, probably in 1963 or 1964.

"The [two-cylinder] tractor was definitely adequate, but some of these legends blow it out of proportion. There's only so much you can do. It had so much to do with the dealer organization, who sold tractors into applications that it could do, and didn't oversell it into too many applications it couldn't do. They knew its limitations so they didn't make stupid moves. You just have to attribute the marketing organization support throughout the U.S. for making it as successful as it was."

SOURCE: Gerald R. Mortensen, Waterloo, Engineering Research, 1954-1964

Paul Champagne's beautifully restored Model 420, photographed at home, spruces up a barn that dates back to the 1930s.

Continued from page 109

The 420W (two-row utility) tractor had a low, wide front and adjustable width. It was the most popular of the tractor-type units. Nearly 11,700 were built in two years. The 420T (row crop) had dual front wheels with narrow spacing. It ranked third when the counting was done, at just over 8,000 tractors. The 420U (utility) was the most common industrial version. It had yellow paint, an adjustable wide front axle, and 17 inches of clearance. The total production was 4,900. A 420C crawler version also was common. There were no mechanical changes from the agricultural version to the industrial, but the industrial version had options to suit the role. The 420S (standard) had an adjustable wide front axle and 22 inches of ground clearance. Dubuque built 3,900 420S tractors.

The 420H (Hi-Crop) was built in a short Phase Three edition 1958 model. The steering wheels on Phase Three machines (except Hi-Crop and LP fuel tractors) were slanted and made of plastic rather than steel. About 610 of the tractors were built. The 420I (special utility) was modified for light industry applications and about 250 models were built.

The 420V (special) was a unique version and less than 90 were built. They would straddle vegetable crops grown on high seedbeds, but were only half as high as the 420 Hi-Crop.

Model 330

The Model 330 is the single most sought-after modern two-cylinder John Deere tractor—period. Similar to the M and 320, it was offered as a low-cost ($2,200) tractor for small farms and little jobs that didn't need much power.

The first Model 330 was marked "built" at the Dubuque factory on August 1, 1958. Two days earlier, NASA had been created as the U.S. responded to the launch of Russia's Sputnik satellite. It was a great year for Hollywood, too. Two of the great classics, *Vertigo* and *Cat on a Hot Tin Roof* were in theaters that year. A teenage alternative, *The Blob*, featured actor Steve McQueen in his first starring role playing a high-school hero who combated a meteorite that oozed a disgusting, gooey substance that ate people.

The Model 330 was the smaller of two 30 Series tractors from Dubuque's factory that were introduced at about the same time. Like the Model 320, it had the basic Model M vertical engine with the 4x4-inch bore and stroke that ran at 1,650 rpm. It achieved a bit more horsepower (21.5 horsepower) on the belt. Mechanically, it was identical to the Phase Two Model 320 with slant dash and steering wheel. Paint-wise, the Model 330 had the bright, streamlined 30 Series paint scheme on sheet metal with fewer creases and bends. It had a lot of options, including an air pre-cleaner, rear exhaust, and an exhaust silencer for use when engine noise was objectionable.

Buyers had a choice of the Model 330 in only two versions, standard or utility. Both were equipped with a four-speed transmission and gasoline engine. The standard was an early precision farming tractor for crops that needed precision work. On a big farm, it became a chore tractor. The utility version was a good, economical, all-purpose tractor.

Both versions are hard to find. Although it was available through February 1960, Dubuque production of the Model 330S (standard) was less than 840 tractors. Sales of the Model 330 utility were even poorer. Dubuque only built about 250 of the 330U. Post-factory, dealerships converted a few 330s into what would have been the Model 330V (special), using parts from a 40V or 420V. A few industrial versions also were built for special order with a red paint scheme. Finally, a few Model 330 tractors were painted bright yellow and sold by the Deere Industrial Division.

Looking back, it can be seen that the Model 330 and Model 430 were a little too close for size and price. For an extra $300, the buyer could purchase a Model 430 with about 20 percent more power. As a result, it outsold the Model 330 by more than twelve to one. Today, the Model 330 is much harder to find and much higher in value, even though the two models are quite similar.

The 330s were small, durable, operator-friendly tractors on small farms. They were relied upon, but weren't abused or abandoned compared to many larger row crop tractors. Today's high value is affected by durability and size as well as scarcity. The bigger the piece of iron, the less likely it will be valued for anything but iron. The Model 330 is a small tractor and about as small as they come for self-propelled field crop machinery. It is easy to store, easy to haul, and relatively easy to handle in a shop.

The Model 330 also is a durable tractor. It didn't need much attention. It did its job and stayed at work on the farm for a long time. They seldom migrated into a fence row or salvage yard. Dubuque built about 1,100 Model 330 tractors in all. It is entirely possible that 800 tractors or more still exist.

Model 430

The first Model 430 tractor was coming down the assembly line in Dubuque about the same time as the Model 330. Three versions were built in the first week of production: Model 430T on July 28, Model 430U on July 31, and Model 430S on August 2, 1958. Meanwhile, over in Waterloo, Deere & Company was readying Models 530, 630, 730, and 830 for production schedules to begin in August.

This 1959 John Deere Model 430T tricycle gas tractor is owned by Verlan Heberer of Belleville, Illinois. This tractor from Heberer's collection reflects the diversity of the five examples in this series from Dubuque. This one has a wide buddy seat.

This 1959 Model 430 W All-Fuel has a striking stance at full axle extension. It's one of only 88 that were built with the All-Fuel option.

Mechanically, the Model 430 was identical to the late slant-steer Model 420. It used the same vertical inline two-cylinder engine with a 7:1 compression ratio and produced 29 horsepower at the belt. It was equipped with a four-speed transmission, although a five-speed could be ordered.

The Model 430 had the bright green and yellow 30 Series design for paint and metal. It also had a few other benefits that would encourage a trade-in from a Model 420. It had some additional instruments, a new more comfortable seat, and the same hydraulic three-point hitch that was used on the Model 320. It also had fender options of regular or heavy duty.

An important option on the Model 430 was the shuttle-type Direction Reverser. It had been introduced on the late 420 and could be installed at the factory or in the field. It permitted the operator to use any forward gear for backing up simply by stopping the tractor, putting in the clutch, and moving a lever forward or backward. For front-end loader work, it was a huge help.

This is a very rare Model 430 Hi-Crop that operates on propane fuel. Dubuque manufactured only five of these tractors. This tractor was restored in 1987. The Model 430HC propane was the feature tractor for the 2008 Two-Cylinder Expo.

The Model 430 chassis became a multipurpose device for many products at the Dubuque factory. This 1960 Model 430F was the basis for the Holt forklift, one of 23 that were tagged as "F" for forklift.

This 1959 Model 430 crawler is equipped for LP gas. In addition to the track undercarriage, it had heavier sheet metal and other components. It has heavy-duty 1/4-inch steel protection for the engine and rubber tracks, which were an option at the time.

In 19 months of production, between July 1958 and February 1960, Dubuque built nearly 15,000 tractors in the 430 Series in seven different versions. A small portion of the 430 fleet used All-Fuel or LP-gas instead of gasoline. Direction Reverser was not available on the LP-gas models or on tractors with a live PTO.

The 430W (row crop utility) was the most popular. Dubuque built nearly 6,000 of these. The Model 430T (tricycle, convertible) was second in production. Deere sold 3,250 of these with a choice of three front ends: single wheel, two-wheel tricycle, or wide front. On the rare side, about 210 were built as Hi-Crop tractors, about 70 were built into forklifts known as the 430U (early) or 430F-3 (late), and another 60-some were built as the 430V (special). The crawler version was popular. Dubuque also built more than 2,000 of the 430C tractors. A special high altitude version was available, so some of these tractors could work at altitudes above 4,000 feet. These engines had a high compression cylinder head to compensate for the reduced air pressure at high elevations.

The Model 430 was a small crawler, but scored big on traction and pulling power. Notice the winch and cable mounted on the back. The Gearomatic winch was factory-approved by Deere & Company.

This 1959 John Deere Model 435 diesel tractor is owned by Lee Pressler of Port Matilda, Pennsylvania. Pressler purchased it from an estate in Louisiana in 1997 and did a complete restoration. Options on this tractor include the swept-back front axle, five-speed transmission, 90 amp battery, and 560 rpm PTO. This tractor was built on September 24, 1959.

Model 435

The Dubuque factory presented a sneak preview of things to come in early 1959 when it introduced the first John Deere Model 435 tractor. It was tagged as completed on January 31. At the time, Walt Disney's *Sleeping Beauty* had just been released and Vince Lombardi was about to sign a five-year contract to coach the Green Bay Packers. CBS-TV had a new hit series with Clint Eastwood in *Rawhide*. Sadly, singer Buddy Holly's final performance was on February 2.

The Model 435 was essentially a Model 430 row crop utility tractor with different footwear and a diesel engine. The John Deere New Generation tractors with four-cylinder engines and many upgrades were to be introduced in 1960. For Deere & Company to meet a niche market need, it was a whole lot more economical

to borrow a good two-cylinder diesel already in production than to develop a whole new engine.

The Model 435 diesel engine was made by General Motors Corporation. It was known as the Detroit 2-53 Diesel engine. It had a 53-cubic-inch displacement for each cylinder, a tremendous compression ratio of 17:1, and a supercharger. It produced almost 33 horsepower on the PTO belt and 28.4 horsepower at the drawbar at 1,850 rpm. It really wasn't a Johnny Popper at all.

Johnny Popper collectors are highly tuned to the sound of the engine. Most collectors have a special love for that slow, deep throaty sound. Some can identify the model by the sound of the engine. The sound of a Model 435 speaks more to the heart of a logger or biker. It has the high-speed sound of a two-stroke engine and fires at every stroke. The new Dubuque tractor, like the early

This 1960 John Deere Model 435 diesel tractor was purchased in 1995 and restored by owner David Frankenfield of Roversford, Pennsylvania. The 435 was manufactured between January 1959 and April 1960. A total of 4,625 were built.

Waterloo Boy Model LA, had evenly spaced power strokes with two horizontally opposed cylinders. True Johnny Poppers have power strokes that cause them to fire on a 180-/540-degree schedule.

The first Model 435 was followed by 4,625 units in the next 13 months. The Farm and Industrial Equipment Institute and the American Society of Agricultural engineers, along with the Society of Automotive Engineers, set up standards later in 1958 for PTOs and hitches. Model 435 became the first John Deere tractor tested by the University of Nebraska under the new 540/1,000 rpm PTO standards.

In North America, starting a Model 435 diesel engine became a problem when the temperature fell below 50 degrees Fahrenheit. Dealerships could fix the problem by adding a little heater into the cooling system. Generally, the Model 435 shared the same chassis and features as the Model 430W.

Compared to a 430T, the new diesel was 21 inches longer, 30 inches wider, 13 inches taller, and weighed an additional 750 pounds. Like the 430, it had a four-speed

transmission with a five-speed option. In fourth gear, it was the fastest two-cylinder ever made by Deere & Company. The speed jumped from 4.75 miles per hour in third gear to 13.5 miles per hour in fourth gear. Maximum pull exceeded 4,200 pounds. Pulling power and weight were about the same as the old AR on distillate tested at Nebraska in 1941.

Deere & Company had an export market for this tractor. It assembled 371 of the Model 435 tractors in Monterrey, Mexico, for sales in the Latin American market. In 1963, the 435 was reincarnated in Rosario, Argentina, as a John Deere Model 445. It stayed in production until 1972.

Model 440

The first John Deere tractor designed from the start for industrial work was the Model 440. It was built as both a crawler and a wheeled, standard-tread tractor from January 1958 through 1960 at the Dubuque factory. This was ahead of the Model 330 and Model 430 at Dubuque, and during the same time that the Model 420 Crawler was still in production. The Model 440 wheel tractors and crawlers were painted industrial yellow. Based on the excellent popularity of the Model 420 series crawlers, Deere & Company managers believed an entirely new series of industrial-purpose machines would be big sellers. They were right. According to www.antiquetractors.com, Dubuque put out 21,928 serial numbers for the Model 440.

From the start, the 440 had industrial styling. It had stronger sheet metal, better radiator protection, and a front-mounted hydraulic pump. For safety, the radiator and fuel caps were located under the hood. The air cleaner was located behind the engine. Like the Slant-dash 420 and 430, it had a slanted instrument panel and comfortable seat.

The final drive housing on Model 420C crawlers had been breaking too often. Deere responded with much stronger casting for the Model 440. Operators could tighten the crawler track with hydraulic power. The four-roller option was cancelled and all 440s were made with five rollers. According to the Model 440I gas crawler operator's manual, it had a capacity to push or pull 6,000 pounds and could handle a 90-inch dozer blade. It had a turning radius of only 78 inches and was only 50.5 inches high at the top of the grille. The shipping weight was 5,850 pounds.

The Model 440I (wheel tractor; gas) and Model 440IC (crawler; gas) were available with the same two-cylinder engine as the Model 420. In 1959, about halfway through the run, the engine was upgraded with a new, higher compression cylinder head. That, plus a boost to 2,000 rpm on the governor, enabled the engine to generate 10 percent more power. Over the two-year production run, approximately 12,200 Model 440 crawlers were sold with the two-cylinder gasoline engine.

The GM 2-53 diesel engine, same as in the Model 435, became an option in 1959 and 1960. About 7,600 Model 440ID (wheel; diesel) and Model 440ICD (crawler; diesel) units were built and sold after the introduction. In addition, Dubuque built an All-Fuel version of the Model 440IC but production stopped at number ten.

Model 435 tractors were noted for the two-cylinder, two-cycle General Motors Detroit diesel engine. In fourth gear, it was the fastest of the John Deere two-cylinder tractors. The 435 was rated for 33 horsepower on the PTO and 29 horsepower on the drawbar. The tractor weighed 3,750 pounds.

At the time, Dubuque was able to offer an innovation called Pilot Touch to crawler operators. A single lever could control both steering and reversing. It was a popular option to order, but it didn't work well. Most crawlers with the Pilot Touch were converted back to the original controls. Other options included a rear PTO, belt pulley attachment, power steering, and direction reverser. The 440 could be used with a front-end loader, backhoe, rear blade, street sweeper, scarifier, snow plow, and three-point hitch implements.

Hidden away in the corn patch, this is a rare 1937 John Deere Model 62 owned by Joseph Nordlander of Van Etten, New York. Nordlander purchased the tractor from the original owner in 1958 and has kept it in continuous regular service for cultivating and side-dressing a few acres of vegetables.

CHAPTER 7

MOLINE FACTORY
BABY BROTHER TRACTORS
MODELS 62, L, LA

The metamorphosis of farming was in full gear when the Great Depression hit. Powerful tractors were breaking the land. Farm wagons and horse-drawn implements were becoming obsolete. In Moline, the John Deere Wagon Works had some extra capacity available. An experimental tractor emerged from the Wagon Works in 1936, followed by a rebuild in 1937. The next year, an unstyled Model L was introduced. It was replaced in 1938 as a styled Model L with a Hercules NXB engine. This was the first John Deere utility tractor that could serve as a primary tractor for small farms and a secondary tractor for larger farms.

Deere & Company managers believed that a Baby Deere utility tractor that was suitable for several hundred thousand small farms would be popular when the economy recovered. This model was about half the size of the Model B and priced at around $500. It could pull a single plow and other implements. The utility tractor went through serious revisions, but it hit the mark. The last tractor was shipped on April 5, 1946 to Omaha. When the shipping doors closed on the last Moline tractor from John Deere, another set of doors at a huge new tractor factory were nearly ready to open in Dubuque, Iowa.

In addition to size, the John Deere utility tractors from Moline are notable for two other reasons. They were the first John Deere tractors to have a vertical two-cylinder engine and they were the first tractors to have a foot clutch. They included the first Number Series tractor from John Deere, the 62, and the only 100th Anniversary tractor for Deere & Company.

Model Y

The engineers and machinists of John Deere Wagon Works Tractor Division had a next-to-nil budget when they began fashioning the littlest tractor. Nordenson had to find an off-the-shelf two-cylinder gas engine for the heart of the new tractor. He selected the Novo C-66 and mounted it on a pair of sturdy, stubby steel tubes. It saved substantial development costs. The Novo C-66 was a stationary version of the back half of a four-cylinder Novo engine. It mounted lengthwise on the tubes, upright, with the rear cylinder sitting over the front wheels. The crankshaft oriented parallel to the travel direction. To this, Nordenson connected a Ford Model A transmission, and he mounted a differential behind that to drive the rear wheels. It became a very short tractor (only 91 inches long) that was highly maneuverable and had excellent visibility. To accommodate components, the driver's position was shifted slightly to the right while the engine and steering were shifted slightly to the left. The steering column and other components also were off-the-shelf Ford Model A components. Other parts came from current Deere products, such as manure spreader wheels.

Users would have liked this experimental tractor. They could shift with the familiar H pattern and use the clutch with their left foot, as they did in cars and trucks. The brakes were a little awkward because each hand had to operate one of the two rear wheel brakes that were located beside the fenders.

It was standard practice to identify experimental tractors at Waterloo with an X before the serial number. Moline used Model Y without explanation. At least 24 (and perhaps 26) Model Y tractors were hand-built in time for early field testing in spring 1936. Sales literature listed the one-plow Model Y at a weight of 1,340 pounds.

The Model Y had issues and good potential. Early on, it was obvious that the 8 horsepower Novo C-66 engine wasn't suitable. Only a few of the Model Y prototypes actually were equipped with this engine. It had barely enough power to move the little tractor. When the right wheels were in the furrow, it didn't have enough oil capacity to plow. That winter, the entire production of Model Y tractors was recalled to the Moline factory. With that, the experimental Model Y disappeared. The last Model Y known to exist was used at Moline to develop a Model 101 rear engine tractor and it never made it into production.

However, thanks to Omaha collector/trucker Jack Kreeger, a Model Y replica exists today in a private collection. Kreeger collected the original literature, consulted with Nordenson, obtained detailed photos, located a few original parts, and had Dan Schmidt (an Omaha machinist-welder) build a new, near-duplicate Model Y.

In a 2007 conversation, Kreeger recalled: "I've got a photograph of the first one they put together, sitting on a wooden dock. That Model Y had quite a few Ford car parts in it. The steering column, the front spindles,

things like that. Now you can look over the fence, where they were taking the pictures, and you can see out in the fenced area there's a junk yard. I told Nordie, I think I know where you got your parts. John Deere said you could build a small tractor if you don't spend any money!"

Model 62

The Model 62 was, in effect, a second stage prototype. It was produced on the assembly line rather than hand-built. According to serial numbers, Moline built 78 of these little tractors. Unofficially, according to Kreeger, the number released from the factory may have been 72 or 73 tractors. They were not recalled for reconstruction, but remained in the field and exist in private collections. Steve Ridenour and sons, specialists in the Model L family at Trenton, Ohio, have listed more than 60 of the original Model 62 tractors in private collections.

The origin of the Model 62 name is uncertain. Most production records were destroyed for both the 62 and the early Model L. The Model 62 was mostly a new tractor, but Nordenson had changed the powertrain. It had a 10-horsepower Hercules NXA engine connected to an integral transmission and rear end. Hercules specialized in small engine development and already was building engines for Deere & Company. Hercules furnished the NXA prototype on short notice by delivering a four-cylinder engine that was cut in half in spring 1936. Two months later, Moline ordered 20 more tractors. Each experimental Model Y was refitted that winter with the NXA and more engines were ordered.

It has been reported that a Spicer transmission and rear end were used with the Model 62. However, information from Spicer records and former employees indicates this was never supplied to Deere & Company. It appears that the Model 62 came out with a John Deere designed-and-built all-in-one transmission-rear end.[13]

Along with the rebuilt tractors, Deere built completely new units as Model 62, starting in March 1937. The Model 62 had many other changes. The three-speed transaxle was a single unit and could operate a belt

[13] Personal recollections of Jack Kreeger, November 2007

This Model 62 is equipped with the only known original Model 62 system for cultivating and side-dressing fertilizer.

pulley to make it more useful on the small farm. It had new sheet metal, including two-piece fenders. It had a cast housing with the letters JD in the mold and carried the serial number plate on the rear axle housing. The front end carried a large casting of the JD logo below the radiator. The first Model 62 had a welded Model Y rear end, but all the rest were cast iron.

Model L Unstyled

Deere Wagon Works was ready to go into full-scale production with its new $500 utility tractor after the summer break in 1937. It had a new tractor that could pull a single 12-inch plow in second gear in most situations. It could handle cultivating 5 to 20 acres very nicely. It was time for a new name that would bring Moline's littlest tractor in line with other Letter Series tractors from Waterloo. It became designated as the Model L. Moline built its first Model L in August 1937 as a 1938 model and rolled 1,500 of these tractors down the line before the 1938 summer shutdown.

At first glance, the unstyled Model L and the former Model 62 were nearly identical. Literature promoted the Model L as a light, economical tractor able to do all the work of a horse or mule without needing a winter feed supply. The Model L had 7 horsepower on the drawbar,

which was just enough to pull that 12-inch plow or another implement. It didn't have a power take-off or hydraulic power to lift an implement, but it did have a foot clutch and an off-set engine to aid visibility while cultivating single rows. Engineers had reduced power requirements, weight, and expense by building it without heavy casting for a main frame. The tractor's basic frame was built up from a pair of sturdy steel tubes. At the dealership, buyers could select from breakaway shovels, regular shovels, V-shaped shovels, and spring-type shovels. Vegetable growers could select from four steel shields that would protect the crop from the cultivator. The tractor could operate two belt pulleys. The primary pulley was side-mounted for operating hammer mills, corn shellers, buzz saws, and other belt-driven equipment. The other pulley was clamped to the driveshaft and exclusively used for operating a mower in small hayfields. For the first time on a John Deere two-cylinder tractor, rubber tires were standard and not optional. Literature offered steel wheels, but there is no record of them being ordered.

The 1937-1938 Model L series continued to use the Hercules NXA vertical two-cylinder engine. For 1939-1940, this change was made to the Hercules NXB. In

Continued on page 124

James Wormley (left) and Ira Maxon pose outside the old Wagon Works factory at Moline with the first Model Y utility tractors built for testing in 1936. Maxon was the project manager, and Wormley went on to become general manager of the new Dubuque factory in 1952. *Jack Kreeger collection*

Willard "Nordy" Nordenson is shown during a test run on Model Y, serial number 101, at Moline. Nordenson is regarded as the father of the Moline tractors. *Jack Kreeger collection*

Management and engineers at Deere & Company were looking to better times and hoped to target at least one untapped tractor market while America struggled through the Great Depression. In mid-1935, the reliable Waterloo Tractor Works factory was tapped out for new capacity. The Waterloo factory was building the great standard-tread Model D, plus two families of row crop tractors. It also was committed to a third and larger row crop tractor family, and there was potential for a fourth family, yet to be named but smaller than the Model B.

At headquarters, it was decided that Deere & Company should attempt to build the smallest-yet farm tractor. Management wanted two-cylinder power in a utility tractor that could appeal to very small farms and part-time farm operators. These were people who needed a horse or small mule to work the ground with a single plow before planting and for weeding. Typically, the small-scale farmer had from 5-20 acres to manage. A true row crop tractor couldn't be justified on a

farm that size, but the farmer would appreciate a small tractor and might sell the horse under the right circumstances.

The project was headed up by three engineers who had almost no budget, a short time-frame for development, and lacked an established tractor factory. The first engineer in Moline was the distinguished senior engineer Max Sklovsky. Sklovsky had joined Deere & Company in 1902 as a rookie in the new field of electrical engineering. Deere & Company had only one factory in Moline, the John Deere Plow Works. His first assignment was to design and supervise the installation of an electrically powered, self-feeding, oil-burning boiler system in the John Deere Plow Works. After earning a master's degree in mechanical engineering, Sklovsky was appointed Chief Engineer in 1910.

Prior to the acquisition of the Waterloo Works, Sklovsky had evaluated farm tractors on the 1912 market. With Dain and other engineers, his efforts to develop a tractor-cultivator were cut short when Deere & Company purchased

The tractor division office at the John Deere Wagon Works in Moline during August 1935. *Jack Kreeger collection*

A 1939 Model L tractor pulls a new John Deere steel wagon with a wooden grain box. Both were built by the John Deere Wagon Works. The Model L sold for $450. This scene is from the King Farm in Moline.

the Waterloo Boy factory. While tractor development proceeded, Sklovsky led numerous engineering tasks in Moline. By 1935, he was chief engineer and had contributed to designing several implements and participated in developing experimental tractors. For instance, he participated in designing the GP, A, and B tractors that were manufactured in Waterloo. He also had patented equipment that increased factory production and reduced costs.

The second engineer was Ira Maxon. He had joined Deere & Company in Moline ten years earlier as a development engineer. He also happened to be the son of Max Sklovsky. At some point, Ira had changed his last name to Maxon.

The third engineer was Willard "Nordy" Nordenson. A specialist in engines, Nordenson had been hired by the Waterloo Works engineering team in 1926. He was laid off and had a short freelance career as an engineer during the Depression. Nordy was rehired at the Moline plant by Ira Maxon in 1935 for the utility tractor project.

Maxon was handed the task of managing this utility tractor project and was assigned space inside the Moline Wagon Works. The Wagon Works plant used to be the former Velie Motors Factory. There was limited space available there for the new project.

The budget was one further catch. Every cent of reserve money had been drained by the combination of Depression-era losses and the recent launch of Models A and B. The research team would have to literally build the new tractor from scratch. The team succeeded with the new product launch in 1938 of the Model L family.

When the Tractor Division was established in Moline during 1941, Ira Maxon was appointed Chief Engineer. In early 1944, Max Sklovsky retired after a 42-year engineering career at Deere & Company. A month later, Ira Maxon left to join with Universal Unit Machinery Co. of Milwaukee. In February 1944, Nordenson was appointed assistant manager of the Tractor Works in Moline. By November, he was promoted to manager.

In time, a patent was issued for the new Model L tractor. Sklovsky, Maxon, and Nordenson were credited as the inventors of the Model L.

Nordenson's achievements were noted and appreciated by Deere President Charles Deere Wiman. Wiman discontinued production of the Model L utility tractor family in April 1946, but he promoted Nordenson to serve as engineering manager for the multi-million-dollar tractor factory that Deere & Company was building in Dubuque. Nordenson was the lead engineer when the Model M was the first tractor out of the new Dubuque factory in 1947. That same year, the Nordenson engineering team moved from Moline to the new engineering quarters in Dubuque. Nordenson, the father of the Models L and LA, retired from Deere & Company in 1968.

In 1946, a young engineer named Mike Mack joined the engineering group at Moline. In 2007, he estimated the Moline factory had employed perhaps 500 workers when the Model L line was in production. Mack transferred to Dubuque in 1947 with Nordenson. In 1956, Mack transferred to Waterloo. He became director of the Project Engineering Center and retired in 1986.

Continued from page 121

1940, Deere & Company was developing its own gas engine for the little tractor family. The company engine, starting in 1941, had a one-piece casting for the block and clutch housing. It also had provisions for a starter and generator.

A three-speed, automotive-style transmission and foot clutch were standard on the Model L. The rear brakes were the same as on the Model 62. They could operate separately and enable the tractor to turn in a seven-foot circle. Adjustable rear tread was available and a big asset to vegetable and row crop growers.

The unstyled Model L also had suffered some cost-cutting. It didn't offer the big JD logo casting below the radiator. Instead, the John Deere name was embossed on the top of the brass radiator tank. At the rear, six-slot steel disks replaced the steel rims with cast centers found on the Model 62. Rear fenders were one-piece construction and a little smaller. They were a bit cheaper to build and made it possible for a buyer to fit a mounted sickle bar mower to the little tractor.

The Model L and its successors also had problems. Front-end spindles and draglinks tended to wear out. The engine was vulnerable to premature wear because the tractor didn't have an oil filter. In fact, the whole series never had an oil filter. If the air cleaner became plugged, the engine could suck in dirt and wear out the rings.

Research indicates that Nordenson and Maxon built one or possibly two tricycle, row crop versions of the Model L equipped with adjustable rear tread. The one known to exist was destroyed, by order of the management, at Dubuque.

Model L Styled

Late in 1937, Henry Dreyfuss and Ira Maxon began corresponding about style changes for next year's Model L. An agreement was reached and a few months later, Moline began tooling up for the 1938 styled Model L. The first styled Model L rolled out the doors on August 15, 1938, powered with the Hercules NXB. This version stayed in production without major further changes until the line closed in 1946. It was a little more powerful and proved

This 1938 unstyled John Deere Model L tractor is owned by Bruce Sniffin of Vernon, Connecticut. It was rebuilt and restored from the ground up and is now in better-than-factory condition.

to be very popular. Deere built and sold nearly 11,000 tractors.

For the styled L, engineers did deepen the NXA cylinder bore in 1940 by 1/4 inch to slightly increase the power. They renamed it the NXB. On these models, the transmission shift lever is located on the right side of the gearbox rather than on top. In July 1941, Moline replaced the Hercules engine with its own John Deere engine. It had the same bore and stroke, same power, same piston rods, same valves, valve springs, and timing gear as the Hercules, but a generator and electric starter could be attached. The clutch housing was cast with the engine rather than bolted to it. Spark plugs were four fingers apart rather than two. It also had a separate set of part numbers. Along with the new engine in 1941, electric lights and an electric starter package became options.

An industrial version of the unstyled L was introduced in 1938 and continued on as a styled L. The number of unstyled Model L industrial tractors is uncertain. Some tractors that came off the line were industrial yellow, but never were recorded differently

This little 1941 styled John Deere Model L tractor is equipped with its own mounted plow. The tractor is owned by Bud Reifsneider of Royersford, Pennsylvania.

from the standard Model L. After 1941, Moline records indicate that 2,018 styled Model LI tractors were built. The Model LI was usually industrial yellow or orange. It had shorter front spindles and rear wheel spacers that gave it a shorter, wider stance. It had the first hydraulics in the L family. With live hydraulic power lift, the LI tractors could operate a No. 7 sickle mower off the driveshaft with a flat belt, an integral gang mower, a belly-mount reel mower, shoulder maintainer, sweep-and-snow brush, and snow plows.

Model LA

The first John Deere Model LA utility tractor was built on August 2, 1940. It was good news in Illinois, but it came among the darkest days of World War II in Europe. One other bright spot was a small article in the *New York Times* that reported that New York's first full-time FM station, W2XOR on 43.4 MHz, was placed in continuous operation as of August 1 for 15 hours daily.

The Model LA was a bigger and better member for the Moline tractor family. The Model LA had 40 percent more power, weighed more, and had more clearance than the Model L. It weighed 2,200 pounds, compared to 1,550 pounds for the Model L at that time. The LA stood two inches taller on 24-inch cast-center rear wheels. It was about the same size as the Farmall Cub tractor, but it was built much heavier. The sturdy steel tubular frame was gone and replaced with solid steel bars. It had adequate power for serious farm work on small farms and it was well built. It could pull a 16-inch bottom plow.

To gain the power increase, the LA engine bore increased to 3.5 inches and the engine speed was bumped from 300 rpm to 1,850 rpm. The Model LA was rated at 13.10 horsepower on the drawbar and 14.34 horsepower on the belt at the Nebraska Test Station in 1941.

Options included an electric starter and generator ($36), lighting ($7), adjustable front axle ($11), a 540-rpm PTO ($30), wheel weights ($7), and a belt pulley ($19). Matched working equipment for the L and LA included one-way and two-way plows, an integral spring-tooth harrow, integral sweep rake, vegetable planter, vegetable cultivator, a series of one-row cultivators with

Here is a side view of a 1941 John Deere Model LA tractor. The tractor's restoration took about six months and required rebuilding the engine, finding replacement wheels and new tires, and much more.

This 1941 John Deere Model LA tractor is owned by Jim and Beverley Kauffman of Fernandina Beach, Florida. This tractor worked in the produce gardens of New Jersey for many years and was fully restored by Kauffman a few years ago.

associated planting and fertilizing equipment, a potato hoe and hiller, an integral side delivery rake, two-row corn planters, a corn drill, two cotton-and-corn planters, a grain drill, field cultivator, tandem disk, single-disk harrow, and other pull-type implements. A PTO was offered on the LA that followed and it was possible to get it dealer-installed on the LI to run off of the driveshaft. A belt pulley was always as an option as well.

From 1941 through 1946, the Moline factory was building three tractors as supplies permitted. Production shut down in 1943 except for 49 Model LI tractors. In 1944 to 1946, Moline built more than 7,000 of the Model LA utility tractors and about 2,500 of the two smaller tractors. The basic model price of Model LA was $560, which was only $65 more than the Baby Deere. Moline built about 12,500 Model LA tractors in five production years. The Moline factory went on to other tasks in 1947. The original tractor factory at Waterloo continued rolling out John Deere's large standard-tread and row crop tractors. By then, the new tractor factory in Dubuque was ready to roll.

BIBLIOGRAPHY

Beemer, Rod, and Chester Peterson Jr. *Inside John Deere: A Factory History*. MBI Publishing Company. St. Paul, MN: 1999.

Beemer, Rod. *John Deere Two-Cylinder Tractors*. MBI Publishing Company. St. Paul, MN: 2003.

Beemer, Rod. *Small John Deere Tractors*. MBI Publishing Company. St. Paul, MN: 2002.

Bollinger, Holly, and Randy Leffingwell. *John Deere Tractors: The First Generation*. MBI Publishing Company. St. Paul, MN: 2004.

Bollinger, Holly. *Classic John Deere Tractors*. Crestline. St. Paul, MN: 2004.

Brown, Theo. *Deere & Company Early Tractor Development*. Two-Cylinder Club. Grundy Center, IA: 1997.

Dietz, John. *John Deere Two-Cylinder Tractor Buyer's Guide*. MBI Publishing Company. St. Paul, MN: 2006.

Dunning, Lorry. *John Deere Tractor Data Book*. MBI Publishing Company. Osceola, WI: 1997.

First Numbered Series of John Deere Tractors. Green Magazine. Bee, NE: 2004.

Hobbs, J. R. *The John Deere Styled Letter Series*. Green Magazine. Bee, NE: 2002.

Hobbs, J. R. *The John Deere Unstyled Letter Series*. Green Magazine. Bee, NE: 2000.

John Deere Tractors: 1918-1994. Deere & Company. Moline, IL: 1994.

Kraushaar, Andy, and Brian Rukes. *Original John Deere Model A*. MBI Publishing Company. St. Paul, MN: 2001.

Leffingwell, Randy. *John Deere: A History of the Tractor*. MBI Publishing Company. St. Paul, MN: 2006.

LeTourneau, P. A. *John Deere Model D Photo Archive*. Iconografix, Inc. Hudson, WI: 1993.

LeTourneau, Peter. *John Deere Limited Production & Experimental Tractors*. MBI Publishing Company. Osceola, WI: 1994.

MacMillan, Don. *John Deere Tractors & Equipment, Vol. 1*. ASAE. Washington, D.C.: 1998.

MacMillan, Don. *The Big Book of John Deere Tractors*. Voyageur Press. Stillwater, MN: 2005.

MacMillan, Don. *The Field Guide to John Deere Tractors*. Voyageur Press. Stillwater, MN: 2002.

MacMillan, Don. *The John Deere Legacy*. Voyageur Press. Stillwater, MN: 2003.

MacMillan, Don. *The Little Book of John Deere*. Voyageur Press. Stillwater, MN: 2003.

Meyer, Faith Hamilton. *John Deere Dubuque Works 1947-1997: Changing Perspectives*. John Deere Dubuque Works. Dubuque, IA: 1997.

Pripps, Robert N. *John Deere GP Tractors: A History in Pictures*. MBI Publishing Company. St. Paul, MN: 2005.

Pripps, Robert N. *John Deere Photographic History*. MBI Publishing Company. Osceola, WI: 1995.

Pripps, Robert N. *Standard Catalog of John Deere Tractors*. KP Books. Iola, WI: 2004.

Rukes, Brian. *John Deere Industrials*. MBI Publishing Company. St. Paul, MN: 2002.

Sanders, Ralph. *Ultimate John Deere*. Voyageur Press. Stillwater, MN: 2001.

INDEX